# SPRINGER TRACTS IN MODERN PHYSICS

Ergebnisse
der exakten Natur-
wissenschaften

Volume **41**

Editor: G. Höhler

Editorial Board: P. Falk-Vairant  S. Flügge  J. Hamilton
F. Hund  H. Lehmann  E. A. Niekisch  W. Paul

Springer-Verlag Berlin Heidelberg GmbH 1966

*Manuscripts for publication should be addressed to:*

G. Höhler, Institut für Theoretische Kernphysik der Technischen Hochschule, 75 Karlsruhe, Kaiserstraße 12

*Proofs and all correspondence concerning papers in the process of publication should be addressed to:*

E. A. Niekisch, Kernforschungsanlage Jülich, Arbeitsgruppe Institut für Technische Physik, 517 Jülich, Postfach 365

ISBN 978-3-662-15921-7     ISBN 978-3-540-34868-9 (eBook)
DOI 10.1007/978-3-540-34868-9

The use of general descriptive names, trade names, trade marks, etc. in this publication, even if the former are not especially identified, is not to be taken as a sign that such names, as understood by the Trade Marks and Merchandise Marks Act, may accordingly be used freely by anyone. Title-No. 4724

# Cluster Representations of Nuclei[*]

## Karl Wildermuth and Walter McClure

## Contents

[*] This work was partly supported by: Bundesministerium für wissenschaftliche Forschung.

# I. Introduction

Numerous experimental and theoretical investigations have confirmed the proposal that nuclei are built up of protons and neutrons, W. HEISENBERG [1] and D. IVANENKO [2]. Therefore it should be possible to derive all properties of a nucleus consisting of $A$ nucleons, or of a nuclear reaction in which $A$ nucleons participate, from the Schrödinger equation for this $A$-body problem.

$$\left\{- \frac{\hbar^2}{2M} \sum_{i=1}^{A} \nabla_i^2 + V(\underset{\sim}{r_1}, \ldots, \underset{\sim}{r_A})\right\} \psi(\underset{\sim}{r_1}, \ldots, \underset{\sim}{r_A}, t)$$
$$= i\,\hbar\, \frac{\partial}{\partial t}\, \psi(\underset{\sim}{r_1}, \ldots, \underset{\sim}{r_A}, t) \tag{I,1}$$

$\underset{\sim}{r_i}$ denotes the space, spin and isobaric spin coordinates of the $i^{th}$ nucleon.

To carry out this program we must overcome two problems:

a) We must know the specific form of the nuclear interaction, $V(\underset{\sim}{r_1}, \ldots, \underset{\sim}{r_A})$.

b) We must solve, at least approximately, the $A$-body Schrödinger equation.

The form of the nuclear potential is in fact not completely known. It has become evident only recently that this potential is primarily a superposition of two-body potentials between all pairs of nucleons, see for instance H. BETHE [3]. The problem with the two-body forces proposed earlier was that they could not produce the nuclear saturation character, e. g. the property that the nuclear volume must increase proportionally with the nuclear weight, $A$. This difficulty led to speculation that the nuclear potential might include many-body forces in addition to the two-body forces. Thus for example, in the case of a three-body problem with an additional three-body force we would have to write

$$V = V_{12} + V_{13} + V_{23} + V_{123}\,.$$

However, over the last fifteen years increasingly reliable two-nucleon potentials have been developed describing the two-nucleon scattering data up to several hundred MeV, see for instance G. BREIT [4]. In order to fit this data at the higher energies the earlier two-body forces had to be modified to include either a hard core or velocity dependent forces. Such improved two-body forces are found capable of producing saturation. Hence it has become gradually clear that potentials without large many-body terms, like $V_{123}$, are adequate to fit the nuclear data, including saturation.

In our choice of a nuclear potential we shall adopt a phenomenological point of view; that is, we shall assume a nuclear potential composed only of two-body forces, and we require only that our two-body potentials be approximately consistent with the two-nucleon scattering data up to two or three hundred MeV, give the correct deuteron binding energy, and have the proper nuclear saturation character, see for instance T. WU and T. OHMURA [5]. The presently unresolved question of which of the possible potentials meeting these criteria is the correct one[*] is more a question for relativistic quantum theory. Hence we write for our $A$-nucleon Schrödinger equation:

$$\left\{ -\frac{\hbar^2}{2M} \sum_{i=1}^{A} \nabla_i^2 + \sum_{i=1}^{A} \sum_{j>i}^{A} V_{ij} \right\} \psi(\underset{\sim}{r_1}, \ldots, \underset{\sim}{r_A}, t)$$

$$= i\,\hbar\,\frac{\partial}{\partial t}\,\psi(\underset{\sim}{r_1}, \ldots, \underset{\sim}{r_A}, t)\,. \tag{I,2}$$

For quantitative calculations a typical two-nucleon potential of the hard-core type meeting our criteria which we shall use is given in equation (VI,2).

Our use of a potential corresponds to forces which act without propagation delays and hence are not relativistically correct. This neglect of relativistic effects and our use of the non-relativistic Schrödinger equation are reasonable for energies less than about 100 MeV per nucleon, since the rest mass of a nucleon is about 940 MeV. With this restriction on the energy we may consider the problem of the nuclear potential, for our purposes, more or less settled; for, whatever the outcome of relativistic theory, it must in the low energy limit yield a potential in reasonable accord with the phenomenological one.

We may now turn our attention to the second problem: we want to develop a flexible but consistent approximation method for solving the $A$-nucleon Schrödinger equation. The need for flexibility arises because the properties of nuclei vary considerably from nucleus to nucleus and even from level to level. Hence we want a method of great generality in which for each individual nuclear state we can, first, systematically use our physical intuition to include every physical effect that might help determine the character of the wave function, and then, quantitatively test and improve this approximate wave function that our intuition has led us to. The ideal quantitative test for this type of approach is the Ritz variational principle; as we shall see in the next section it can accomodate and improve the most general sort of trial wave function.

The way we first employ our intuition in an orderly yet flexible manner to arrive at good trial wave functions for the Ritz principle, with a few significant variational parameters, constitutes the method of cluster representations. We shall see that cluster representations when combined with the Ritz variational principle give us a powerful method to solve approximately the $A$-nucleon Schrödinger equation.

Our program is now as follows: after a discussion of the Ritz variational principle, we shall introduce the cluster method using frequently examples of actual nuclei. With the aid of this method we shall try to consider nuclear reactions and nuclear structures from a unified point of view. From this viewpoint, and with the help of the Pauli principle, we shall also be able to understand the relations between the various nuclear models currently in use.

---

[*] It turns out that in fact the wave function is usually relatively insensitive to the exact form of phenomenological potentials meeting the above criteria, see for instance [73], [89].

## II. The Ritz Variational Principle

### A. The Ritz variational procedure

Most of the approximation methods commonly employed to solve the Schrödinger equation, such as the Born approximation or standard perturbation methods, depend on certain arbitrary, and often unknown, parameters being small. This greatly restricts our freedom in trying to set up approximate solutions. On the other hand the Ritz variational procedure not only gives us a quantitative measure of the goodness of absolutely any trial solution we might propose, it also gives us a way to improve this trial solution to the maximum extent of whatever variational parameters we have put in it.

We shall start with a brief review of the Ritz variational principle. Let a wave function $\psi$ be determined by the requirement that the variation of $\int \psi^* H \psi \, d\tau$ vanishes, subject to the constraint of normalization $\int \psi^* \psi \, d\tau = 1$. Introducing the constraint by means of a Lagrange multiplier, we have the equation

$$\delta \left\{ \int \psi^* H \psi \, d\tau + \lambda \int \psi^* \psi \, d\tau \right\} = 0 . \tag{II,1}$$

$H$ is the Hamiltonian of our quantum mechanical system; $\int d\tau$ corresponds to integration and summation over all continuous and discrete coordinate variables.

If we write

$$\psi = \psi_1 + i \, \psi_2 \tag{II,2}$$

with $\psi_1$ and $\psi_2$ both real functions then the expression above must vanish as either $\psi_1$ or $\psi_2$ is varied independently. It can be seen immediately that the same final result is obtained if we instead consider that $\psi^*$ and $\psi$ are varied independently. Varying $\psi^*$ gives

$$\int \delta \, \psi^* (H \, \psi + \lambda \, \psi) \, d\tau = 0 \tag{II,3}$$

and since the value of $\delta \, \psi^*$ at each point is arbitrary, this implies

$$H \, \psi + \lambda \, \psi = 0 \tag{II,4}$$

which is the Schrödinger equation if we identify $\lambda = -E$. Thus requiring that $\psi$ satisfy the above variational requirement is equivalent to requiring that it satisfy the Schrödinger equation.

Let the correct eigenfunctions for our Hamiltonian be, in order of increasing energy, $\phi_0, \phi_1, \ldots, \phi_i$ corresponding to energies $E_0, E_1, \ldots, E_i$. These eigenfunctions form a complete set in terms of which any function, and hence any trial solution $\psi(\underset{\sim}{r_1}, \ldots, \underset{\sim}{r_A})$ can be expanded:

$$\psi(\underset{\sim}{r_1}, \ldots, \underset{\sim}{r_A}) = \sum_{i=0}^{\infty} a_i \, \phi_i(\underset{\sim}{r_1}, \ldots, \underset{\sim}{r_A}) . \tag{II,5}$$

Hence

$$\langle H \rangle \equiv \int \psi^* H \psi \, d\tau = \sum_i |a_i|^2 E_i \geq E_0 \sum_i |a_i|^2 = E_0 . \tag{II,6}$$

This shows that $\int \psi^* H \psi \, d\tau \geq E_0$ for any trial function so that $\int \psi^* H \psi \, d\tau$ corresponds to an absolute minimum if $\psi$ is the ground state eigenfunction $\phi_0$, but only to a locally stationary value if $\psi$ is an excited state. It is easy to see that if we add a second constraint to the test function, $\int \phi_0^* \psi \, d\tau = 0$ the absolute minimum will now occur when $\psi$ is the eigenfunction of the first excited state. Similarly we can handle higher states by imposing constraints of orthogonality to all preceding states.

It is important to note that the magnitude of $\langle H \rangle$ in (II,6) depends on the absolute squares of the expansion coefficients $a_i$. Hence $\langle H \rangle$ can approach the ground state energy $E_0$ very closely only if the absolute values of the $a_i$ for the excited states become small. By (II,5) this means that correspondingly the trial wave function $\psi$ must approach the ground state function $\phi_0$. Therefore the difference between the value of $\langle H \rangle$ and any lower bound* on $E_0$ provides a quantitative test of the goodness of our approximate solution, a test which becomes the more sensitive the closer $\langle H \rangle$ approaches $E_0$.

Let us now summarize our procedure for obtaining approximate eigenfunctions of our Hamiltonian. Let $\psi(\alpha_1, \ldots, \alpha_N, \underset{\sim}{r_1}, \ldots, \underset{\sim}{r_A})$ be a normalized trial wave function, depending upon $N$ adjustable parameters, $\alpha_1, \ldots, \alpha_N$ as well as the dynamical variables $\underset{\sim}{r_1}, \ldots, \underset{\sim}{r_A}$, and of such a form that it may be expected on physical grounds to give a good approximation to the ground state for correctly chosen values of $\alpha_1, \ldots, \alpha_N$. We choose those values $\bar{\alpha}_1, \ldots, \bar{\alpha}_N$ which make $\langle H \rangle$ an absolute minimum; the form of $\psi$ corresponding to these particular values is denoted by $\bar{\psi}_0$ and is our approximate solution for $\phi_0$. Now we choose a new normalized test function and again minimize $\langle H \rangle$ subject to the additional constraint $\int \bar{\psi}_0^* \psi \, d\tau = 0$, thus obtaining $\psi_1$, our approximate solution for $\phi_1$; and we continue to work upward in this manner. We see from this that the quality of an approximate wave function for an excited state usually depends on the quality of the approximate wave functions for the lower excited states.

Usually we will further restrict our trial wave function by requiring that it have specified values for such quantum numbers as the angular momentum and the parity. Trial functions, eigenfunctions, and approximate solutions corresponding to different sets of such quantum numbers will be automatically orthogonal to each other. The computation of the eigenfunctions for various energies corresponding to a given set of quantum numbers can be carried out by the above scheme; the approximate solutions obtained for one such set will not depend in any way upon those obtained for another set.

---

* Just as $\langle H \rangle$ has been found to be an upper bound on $E_0$, so we can find other expressions which provide a lower bound. In fact this is necessary in practice as one means of checking how good a minimum of $\langle H \rangle$ we have obtained. For information on lower bounds, see for instance E. SCHMID [6]. Another practical test of the goodness of the minimum of $\langle H \rangle$ is to show that it is not appreciably lowered by the addition of new variational parameters.

## B. Considerations on trial wave functions

It is the purpose of the cluster method to arrive at good trial wave functions $\psi(\alpha_1, \ldots, \alpha_N, \underset{\sim}{r_1}, \ldots, \underset{\sim}{r_A})$ for the start of the Ritz variational calculations. We want here to make some general remarks on finding trial functions. Suppose we have any complete set $\{u_i\}$. Then we can expand our trial function as a superposition of these:

$$\psi = \sum_{i=0}^{n} b_i u_i(\underset{\sim}{r_1}, \ldots, \underset{\sim}{r_A}) . \tag{II,7}$$

Now let us regard the expansion amplitudes $b_i$ as variational parameters.

If we take only the first term in the summation, we will simply obtain $\psi = u_0$. As $n$ is increased the new parameters $b_1$, $b_2$, $b_3$, etc. are introduced, the trial wave function becomes more flexible, and our solution will correspondingly improve. In the hypothetical limit as $n$ approaches infinity, our trial wave function will be able to take any form and our solution therefore will be exact. But this knowledge provides us scant comfort since in practice we can calculate with only a small number of terms.

Unless one of the functions in our expansion, $u_0$ say, is close to the correct solution, $\phi$, any improvement in our trial function is going to require so many terms and variational parameters as to make the calculation impractical. A suitable expansion system must give a good approximate solution, in the frame of the Ritz variational principle, with only a few terms*.

Because the properties of nuclei can differ so greatly from one nuclear state to the next, we must expect that their wave functions will also differ considerably. Hence, as we shall see, the most suitable expansion system for our trial wave function will vary from nucleus to nucleus and even from level to level. One main purpose of the cluster method is to provide the most appropriate expansion system for each individual nuclear state.

After obtaining a suitable expansion system, we can further improve our calculation by introducing internal parameters into the expansion functions themselves,

$$u_i(\underset{\sim}{r_1}, \ldots, \underset{\sim}{r_A}) \to u_i(\alpha_1, \ldots, \alpha_N; \underset{\sim}{r_1}, \ldots, \underset{\sim}{r_A}) .$$

By the addition of such new variational parameters it is possible to include every physical effect which might help to lower the energy expectation value for a given nuclear state.

---

\* The following non-rigorous condition may give us some insight as to when we will need only a few terms in the expansion for $\psi$. If we use the ansatz (II, 7) with $n$ set equal to infinity, and vary the amplitudes $b_i$ to make $\langle H \rangle$ a minimum, then the following relation must hold, assuming for simplicity the $\{u_i\}$ orthogonal:

$$b_i = \left( \sum_{i \neq j}^{\infty} b_j \langle u_i |H| u_j \rangle \right) \Big/ (\langle u_i |H| u_i \rangle - E_0) .$$

If we wish to neglect the term $b_s u_s$ in the expansion for $\psi$, we see qualitatively from this that $b_s$ will be small if the energy difference $\langle u_s |H| u_s \rangle - E_0$ is large and if simultaneously the overlap integral $\langle u_s | H | u_j \rangle$ is small for all $j \neq s$.

Certainly one could in principle stick to a single complete eigen-function system, as for instance the shell model system, to describe the behaviour of all nuclei. But in practice unless one uses the most suitable expansion system for a given problem the physical insight quickly gets lost and very often the calculation become impossibly difficult. The situation here is similar to that in analytic geometry where a most conve-nient coordinate system usually exists for a particular problem.

Finally we must emphasize that choosing trial functions is in the end a trial and error process. The cluster method, which we shall introduce in detail in the next section, is not a calculating machine; it simply enables us to formulate trial functions systematically incorporating all our useful physical knowledge. Which cluster representation we shall employ and which internal parameters we shall vary is a matter of judgement, and we must finally quantitatively ascertain with the Ritz principle the importance of each physical effect in providing a better approximation to our $A$-body Schrödinger equation.

## III. Oscillator Cluster Representation

We now begin to introduce the method of cluster representations in detail. In this section we will show how our knowledge of energetically preferred substructures in nuclei can lead us to very useful expansion systems, $\{u_i\}$, for our antisymmetrized nuclear trial wave functions. The members of these complete sets we shall call cluster functions. In quantitative calculations with these preliminary cluster functions we shall use rather simple two-nucleon forces which do not include a hard-core force. In a later section we will show step by step how the individual cluster functions can be generalized by the introduction of internal parameters to include hard-core forces and every other significant effect. We will discuss carefully the physical consequences of introducing each new parameter. In the end we shall see that, when properly chosen, very often only one or a few cluster functions need to be used in our Ritz variational principle to obtain a good approximate solution of the $A$-nucleon Schrödinger equation (I,2).

### A. What is a cluster

Because the Schrödinger equation is so mathematically difficult to solve, even approximately, we must employ our physical knowledge as much as possible in the construction of trial nuclear wave functions. Two physical features predominate in determining the wave function: the character of the nuclear forces, and the effects of the Pauli exclusion principle in systems of nuclear dimensions. The obvious facts about the phenomenological nuclear forces are that they are short-ranged ($\approx 10^{-13}$ cm), are strongly attractive over most of this range, but at extreme short distances ($<0.4 \cdot 10^{-13}$ cm) become strongly repulsive (the so-called hard core). The effect of the Pauli principle in a system of nuclear dimensions is to allow low energy nucleons to move relatively freely throughout the nuclear volume, because they may not be scattered into other occupied levels.

These properties severely restrict the motions of the nucleons. In order to minimize the total energy they must take maximum advantage of the attractive potential energy compatible with these restrictions. If the nucleons are too far apart they lose attractive potential energy and the total energy rises. If the nucleons are too close together, kinetic energy effects (from both the uncertainty principle and the Pauli principle) and the repulsive core again raise the total energy. Hence the nucleons must spend a good deal of time (more accurately, there must be a high probability) such that their separation distances lie within this narrow energetically favoured range. But again the Pauli principle forbids any very stringent correlations to achieve this.

There remains a type of loose but effective correlation by which the total energy can be brought to a minimum. We can begin to understand this correlation with a simplified physical argument which we subsequently refine. Consider four nucleons moving in the volume of a larger nucleus. For the moment let us suppose the effect of all other nucleons in this nucleus is to confine these four to the nuclear volume; we also assume temporarily that the Pauli principle is inoperative. Suppose that the four move completely uncorrelated from one another. Then most of the time they will be outside the range of each others attractive forces and there will be little mutual potential energy among them. Now, on the contrary, suppose we introduce a correlation such that there exists the probability that they will be found, more often than not, close to one another. As this probability is increased, within the above mentioned limits, the attractive potential energy among the four nucleons can be greatly increased without a particularly large increase in kinetic energy, and thereby the total energy of the nucleus is reduced.

This simple density correlation is greatly modified by the Pauli principle, however, which we must now add into our picture. Because nucleons are indistinguishable, any of these four nucleons can exchange places with any other nucleon in the nucleus. Thus any individual nucleus can move throughout the nuclear volume while at the same time somewhere in the nucleus four of the nucleons are usually participating in the correlation. Second, unless the spins and isospins of these four nucleons are anti-alligned, the Pauli principle forbids them to be brought very closely together. More drastically, the presence of the other nucleons means that many single particle levels of the nucleus are already occupied. Hence there is much less freedom to correlate the low energy nucleons (which requires a superposition of single particle levels)*; they spread out more through the nuclear volume and can participate

---

* Any nuclear wave function may be arbitrarily expanded in products of single particle functions, although these are not necessarily the most convenient representations.

only weakly in the correlation*. This sort of correlation, where within the above restrictions of the Pauli principle there exists an enhanced probability to find four nucleons, with properly alligned spins and isospins, more often than not especially close together, we call an α-cluster.

We can immediately generalize the idea. Consider $Li^6$ as an example. In the ground state we should expect four of the (indistinguishable) nucleons to correlate as a α-cluster. However, because the Pauli principle prevents more than four nucleons from entering into the mutual short-ranged interaction region, we expect the remaining two to correlate most strongly with each other, forming a deuteron cluster. The structure of $Li^6$ is then mainly described by these two clusters, roughly speaking, moving about each other. In heavier nuclei correlations can exist between smaller clusters so that even larger clusters are built up. The cluster structure of a given nucleus can change with increasing excitation energy due to the changed energetical and orthogonality requirements. Because the Pauli principle drastically limits the variety of structures possible in the nucleus these changes in cluster structure are not liable to differ too greatly from one another, although certain sensitive processes (gamma transitions, level widths, etc.) may be much altered thereby.

The theoretical justification of the above picture rests in formulating wave functions embodying cluster correlations and demonstrating with the Ritz principle that these "cluster functions" provide good approximate solutions to the nuclear Schrödinger equation**. We now turn to the problem of formulating such cluster functions quantitatively.

## B. Construction of oscillator cluster wave functions

It is well known that nuclei exhibit different kinds of behaviour. Some of these are due to single particle features, others are connected with collective motions of the nucleons. The relative importance of the various types of behaviour can change significantly from nucleus to nucleus and even from one level to the next. Thus there usually exists a particular set of single-particle or collective coordinates describing a given nuclear level most simply and clearly.

It is very advantageous to find expansion systems for the nuclear wave function which employ this most appropriate coordinate set from the beginning. We have already observed that the oscillator potential

---

* This means the weak correlated substructure described above can not be naively considered as an α-particle within the nucleus. A partial exception occurs in configurations where the four nucleons are somewhat separated from the others, in which case they can act somewhat more like a free α-particle. Hence we expect the correlation to be strongest near the nuclear surface.

** Since a cluster correlation represents only a probability statement about certain nuclear configurations, there is no direct experimental way to measure a cluster anymore than there is a direct way to measure the wave function itself. However we can certainly predict experimental quantities with these wave functions.

generates a complete set of functions, the common shell model eigen-function system, which can be used to expand a trial wave function for our phenomenological nuclear Schrödinger equation, (I,2), see for instance S. MOSZKOWSKI et al. [7]. But the shell model system is only of limited practical value in expanding trial functions, because it is restricted to single particle coordinates. We now want to investigate the consequences of introducing various sets of collective coordinates into the oscillator Hamiltonian in order to generate new, more practical expansion systems for solving the nuclear Schrödinger equation in the frame of the Ritz variational principle.

There are several reasons why we start with the oscillator potential. First, this potential reflects many broad features of the actual averaged nuclear potential, yet even upon introducing collective coordinates it is simple enough to solve exactly. Second, in order to see that the effects of antisymmetrization serve to remove the contradictions among the different collective and single particle viewpoints, it is important that we be able to compare our new expansion systems with each other and with the shell model system; this is particularly easy if we use the oscillator potential to generate all the systems. Finally, the functions in these new oscillator expansion systems can be generalized by the intro-duction of internal parameters in a very natural way to obtain good trial functions for the Ritz variational principle of our actual Hamiltonian.

We shall first solve the oscillator Hamiltonian using single-particle coordinates, and then show how to introduce collective coordinate sets to find new expansion systems, K. WILDERMUTH and TH. KANELLO-POULOS [8]. Consider the motion of $A$ nucleons in an oscillator potential. Their Schrödinger equation is

$$\frac{1}{2M}\left\{\sum_{k=1}^{A} p_k^2 + a^2\hbar^2 \sum_{k=1}^{A} r_k^2\right\} \psi_n(r_1 \dots, r_A) = E_n \psi_n(r_1, \dots, r_A) \quad (III,1)$$

where $p_k = \frac{\hbar}{i} \nabla_k$ is the operator of the momentum vector of the $k^{th}$ nucleon and $r_k$ is the position vector of the $k^{th}$ nucleon. $a = M\omega/\hbar$ is the width parameter of the oscillator potential. $M$ is the mass of a nucleon and $\omega$ is the angular frequency of the oscillator potential.

The eigenfunctions $\psi_n(r_1, \dots, r_A)$ solving equation (III,1) are pro-ducts of oscillator functions and form a complete set of orthogonal functions for arbitrary wave functions of $A$ variables, $r_k$. We can express the antisymmetrized wave function of any state of a nucleus composed of $A$ nucleons as a superposition of these eigenfunctions, if in addition we introduce the spin and isobaric spin functions of the nucleons. This single particle expansion system usually has limited utility for trial functions because many terms are required to adequately represent any collective behaviour of the nucleons.

Now, collective motion occurs when a certain number of nucleons are energetically favoured to move more or less together; there may be one or more such subgroups within a nucleus. Therefore to introduce more effective coordinates, let us divide the $A$ nucleons into $l$ groups, or clusters$^\star$ as we shall call them, the $k^{\text{th}}$ cluster consisting of $n_k$ nucleons such that $\sum\limits_{k=1}^{l} n_k = A$. The set of indices $\{n_1, n_2, \ldots .n_l\}$ denote what we call a cluster representation of the nucleus. If we now introduce the center of mass coordinates $\boldsymbol{R}_k$ and the center of mass momenta $\boldsymbol{P}_k$ of these clusters, the oscillator model Schrödinger equation (III,1) becomes

$$\left\{ \sum_{k=1}^{l} \left( H_k + \frac{1}{2Mn_k} \boldsymbol{P}^2 + \frac{a^2 \hbar^2 n_k}{2M} \boldsymbol{R}_k^2 \right) \right\} \psi_n = E_n \psi_n \qquad \text{(III,2)}$$

$$\boldsymbol{P}_k \equiv \sum_{s=1}^{n_k} \boldsymbol{p}_s; \quad \boldsymbol{R}_k = \frac{1}{n_k} \sum_{s=1}^{n_k} \boldsymbol{r}_s \qquad \text{(III,2a)}$$

The Hamiltonian $H_k$ depends only on the relative coordinates $\boldsymbol{r}_s - \boldsymbol{R}_k$, and their derivatives, of the nucleons in the $k^{th}$ cluster. The eigenfunctions of (III,2) are therefore products of functions in which every function either depends only on the relative coordinates of one cluster or depends on a single centre of mass coordinate $\boldsymbol{R}_k$.

After introducing the spin and isobaric spin coordinates and anti-symmetrizing, all these eigenfunctions systems are equivalent to the antisymmetrized single particle wave function system — the usual oscillator shell model system — in the sense that we can expand all the states of a nucleus in terms of any one of these eigenfunctions systems. Therefore each cluster expansion system is complete; note however that degenerate eigenfunctions within a cluster system are not necessarily orthogonal. Thus we see that to each cluster representation there belongs an antisymmetrized complete oscillator eigenfunction system, which we can characterize by the representation indices $\{n_1, n_2, \ldots, n_l\}$.

We might also note another important point here that because we have only performed a mathematical transformation on our original oscillator Hamiltonian all these eigenfunction systems must have identically the same energy spectrum. This means that any eigenfunction corresponding to a given energy eigenvalue in one eigenfunction system can be expanded in any other eigenfunction system as a linear super-position of just those degenerate eigenfunctions corresponding to the same energy eigenvalue. This makes the comparison between eigenfunctions of different oscillator eigenfunction systems especially simple. We shall see the utility of such a comparison later.

It now depends on the nuclear forces in which eigenfunction system the different states of the nuclei are most simply represented, i.e. which kind of correlation between the nucleons is particularly favoured in the different nuclear states. This favoured representation, as we have emphasized, can change from nucleus to nucleus and from level to level. One practical criterion for what constitutes a simple description of a state in a certain eigenfunction system is that in this expansion system

---

$^\star$ We must be careful not to think of such substructures too literally as composite "real particles"; exchange effects due to the Pauli principle complicate any such simple, intuitive picture.

the state is described essentially by one or by a superposition of only a small number of antisymmetrical eigenfunctions.

*Antisymmetrized* eigenfunctions of the kind discussed above are called oscillator cluster functions. It must be emphasized that the anti-symmetrization very often influences the physical properties of a wave function profoundly*. Therefore one has always to investigate in principle how much of the physical properties of an unsymmetrized cluster wave function remain after antisymmetrization. This important point will be discussed later in great detail. Special oscillator cluster functions have been given by J. ELLIOT and T. SKYRME [9] and D. BRINK [10].

## C. Discussion of Be⁸ as an illustrative example

The best way to understand the physical significance of the oscillator cluster functions is to consider a specific example. Let us therefore apply the general transformations discussed above to an actual case, the ground state and lowest excited states of Be⁸, K. WILDERMUTH and TH. KANELLOPOULOS [8]. We begin by using our physical knowledge to decide what cluster representation we shall use.

A cluster consisting of two protons and two neutrons and corresponding before antisymmetrization of the nuclear wave function to an α-particle in the ground state we shall call an α-cluster. Because of the high binding energy of a free α-particle, we can presume that the α-cluster is a rather stable substructure particularly in light nuclei where, aside from exchange effects, little distortion should occur. Hence we shall assume the Be⁸ nucleus can be represented to a good approximation by the motion of two α-clusters, with the low excited states corresponding to excitations in the relative motion of the two clusters but with no internal excitation of the clusters themselves**.

We must now test this picture quantitatively. We generate suitable variational trial wave functions by transforming the single particle oscillator Schrödinger equation:

$$\left\{ \frac{1}{2M} \sum_{k=1}^{8} (p_k^2 + a^2\hbar^2 r_k^2) \right\} \psi_n (\underset{\sim}{r_1}, \ldots, \underset{\sim}{r_8})$$

$$= E_n \, \psi_n (\underset{\sim}{r_1}, \ldots \underset{\sim}{r_8}) \tag{III,3}$$

where $\underset{\sim}{r_k}$ denotes space $r_k$, spin $s_k$ and isobaric spin $t_k$ coordinates of the $k^{\text{th}}$ nucleon.

Because they will shortly be useful for comparative purposes, let us first explicity write the single particle solutions of (III,3):

$$\psi_n (\underset{\sim}{r_1}, \ldots \underset{\sim}{r_8}) = \psi_i (\underset{\sim}{r_1}) \, \psi_j (\underset{\sim}{r_2}) \ldots \psi_k (\underset{\sim}{r_8})$$

---

* We want to emphasize that antisymmetrization of the nuclear wave functions is the quantitative expression of the fact that nucleons are indistinguishable and obey the Pauli exclusion principle.

** Such levels we call α-particle states. However, see our cautionary remarks in footnote of p. 10.

where each $\psi_j\,(r_k)$ is an oscillator function multiplied by the appropriate spin and isobaric spin functions (see text to equation III,1). The solutions $\psi_n\,(r_1,\ldots r_8)$ are complete and give rise to a complete set of anti-symmetrized functions, in terms of which any antisymmetrized 8-nucleon wave function can be expressed. We write for the antisymmetrized wave functions

$$\mathscr{A}\;\psi_n\,(r_1,\ldots r_8) = \sum_{P_k}\;(-1)^{P_k}\,P_k\,\psi_n\,(r_1,\ldots r_8) \qquad \text{(III,4)}$$

In eq. (III,4) the sum goes over all different even and odd permutations $P_k$ which can be carried out on the coordinates $r_1,\ldots r_8$.

We introduce now new coordinates appropriate to the cluster view-point:

$$R_1 = \frac{r_1 + r_2 + r_3 + r_4}{4}\;;\; \textit{center of mass coordinate of cluster 1}$$
$$\text{(III,5)}$$
$$R_2 = \frac{r_5 + r_6 + r_7 + r_8}{4}\;;\; \textit{center of mass coordinate of cluster 2}$$

$$\begin{cases}\bar{r}_1 = r_1 - R_1 \\ \bar{r}_2 = r_2 - R_1 \\ \bar{r}_3 = r_3 - R_1 \\ \bar{r}_4 = r_4 - R_1 = -(\bar{r}_1 + \bar{r}_2 + \bar{r}_3)\end{cases} \begin{cases}\bar{r}_5 = r_5 - R_2 \\ \bar{r}_6 = r_6 - R_2 \\ \bar{r}_7 = r_7 - R_2 \\ \bar{r}_8 = r_8 - R_2 = -(\bar{r}_5 + \bar{r}_6 + \bar{r}_7)\end{cases} \qquad \text{(III,6)}$$

In eqs. (III,6) are the respective internal coordinates of the two clusters. Note that not all the $\bar{r}_s$ are independent since they always satisfy $\sum_{s=1}^{4}\bar{r}_s = 0$ and $\sum_{s=5}^{8}\bar{r}_s = 0$. For definiteness we shall say that $\bar{r}_4$ and $\bar{r}_8$ are dependent and the rest independent as already indicated in (III,6). Introducing the new coordinates into the Hamiltonian (III,3) gives

$$H = H_1 + H_2 + \frac{1}{2M}\frac{1}{4}\{P_1^2 + P_2^2 + a^2\hbar^2 4\,(R_1^2 + R_2^2)\} \qquad \text{(III,7)}$$

where $$P_1 \equiv \frac{\hbar}{i}\,\text{grad}_{R_1} \quad \text{and} \quad P_2 \equiv \frac{\hbar}{i}\,\text{grad}_{R_2}$$

are the operators of the total momenta of the clusters *1* and *2*. $H_1$ and $H_2$ are Hamiltonians having as variables the internal coordinates of cluster *1* and cluster *2* respectively, explicitly $H_1$ is given by

$$H_1 = \frac{\hbar^2}{2M}\sum_{j=1}^{3}\sum_{k=1}^{3}\left(-\left(\delta_{jk} - \frac{1}{4}\right)\nabla_{\bar{r}_j}\cdot\nabla_{\bar{r}_k} + a^2(\delta_{jk} + 1)\bar{r}_j\cdot\bar{r}_k\right) \qquad \text{(III,8)}$$

$H_1$ and $H_2$ may be shown to have harmonic oscillator form when expressed in terms of a properly chosen set of three new independent coordinates. This procedure is explicitly carried out in appendix A.

Because of the separability of the Hamiltonian (III,7), its eigenfunctions may be expressed as products:

$$\psi_n\,(R_1, R_2, \bar{r}_1,\ldots\bar{r}_8, s_1,\ldots t_8)$$
$$= \phi_i\,(\bar{r}_1,\ldots\bar{r}_4)\,\phi_j\,(\bar{r}_5,\ldots\bar{r}_8)\,\chi_k\,(R_1)\,\chi_l\,(R_2)\,\xi_m\,(s_1,\ldots t_8) \qquad \text{(III,9)}$$

The total spin and isobaric spin dependence of $\psi_n\,(\boldsymbol{R}_1, \ldots t_8)$ is contained in $\xi\,(s_1, \ldots s_8, t_1, \ldots t_8)$ .

We make one further simple coordinate change:

$$\boldsymbol{R}_{\mathrm{CM}} = \frac{1}{2}\,(\boldsymbol{R}_1 + \boldsymbol{R}_2) = \text{\textit{center of mass coordinates of the total}}$$
$$\text{\textit{8-nucleon system}} \qquad \text{(III,10)}$$
$$\boldsymbol{R} \;\;= \boldsymbol{R}_1 - \boldsymbol{R}_2 \;\;= \text{\textit{relative coordinates of the two clusters}}$$

The Hamiltonian again is separable and of harmonic oscillator form with respect to the two new variables; therefore we can write:

$$\psi_n\,(\boldsymbol{R}_{\mathrm{CM}},\,\boldsymbol{R},\,\bar{r}_1, \ldots t_8)$$
$$= \phi_i\,(\bar{r}_1, \ldots \bar{r}_4)\,\phi_j\,(\bar{r}_5, \ldots \bar{r}_8)\,\chi_k\,(\boldsymbol{R})\,\xi_l\,(s_1, \ldots t_8)\,Z_m\,(\boldsymbol{R}_{\mathrm{CM}}) \qquad \text{(III,11)}$$

The last factor, giving an oscillatory motion of the center of mass of Be⁸, arises because the harmonic oscillator potential is fixed in space and not invariant with respect to center of mass translation, whereas the actual nuclear potential clearly is translation invariant. Hence, we usually drop the $Z_m\,(\boldsymbol{R}_{\mathrm{CM}})$ factor★ and write:

$$\psi_n\,(\boldsymbol{R},\,r_1, \ldots t_8) = \phi_i\,(\bar{r}_1, \ldots \bar{r}_4)\,\phi_j\,(\bar{r}_5, \ldots \bar{r}_8)\,\chi_k\,(\boldsymbol{R})\,\xi\,(s_1, \ldots t_8) \qquad \text{(III,12)}$$

Formula (III,12) is the general form of a two cluster wave function. For the sake of general terminology we call $\phi_i\,(1)$ and $\phi_j\,(2)$ the internal functions of cluster 1 and cluster 2 respectively, and $\chi_k\,(\boldsymbol{R})$ the relative motion function.

Again the $\psi_n\,(\boldsymbol{R},\,\bar{r}_1, \ldots t_8)$ are a complete set for nuclear wave functions and the $\mathscr{A}\,\psi_n\,(\boldsymbol{R}, \ldots t_8)$ are a complete set of antisymmetrized functions (oscillator cluster functions). Now we introduce a restriction by requiring that $\phi_i\,(\bar{r}_1, \ldots \bar{r}_4)$ and $\phi_j\,(\bar{r}_5, \ldots \bar{r}_8)$ be in their ground states, $\phi_0\,(\bar{r}_1, \ldots \bar{r}_4)$ and $\phi_0\,(\bar{r}_5, \ldots \bar{r}_8)$, and that the spin and isobaric spin variables of the nucleons in each cluster cancel each other to $S = 0$ and $T = 0$. This restriction we assume from our physical knowledge that breakup of the $\alpha$-cluster substructures inside the nucleus is very unfavourable energetically.

Since all we have done so far is to change coordinates, the Hamiltonians in eq. (III,3) and (III,1) are physically identical and must therefore have identical energy eigenvalues. Hence (see text p. 10) each oscillator cluster function of a given energy may be expressed as a superposition of degenerate single-particle solutions (III,4) of that same energy.★★

$$\mathscr{A}\,\psi_n\,(\boldsymbol{R}_{\mathrm{CM}},\,\boldsymbol{R},\,\bar{r}_1, \ldots t_8) = \sum_m \mathscr{A}\,\psi_m\,(r_1, \ldots r_8) \qquad \text{(III, 12a)}$$

---

★ The possibility of unambiguously identifying a center of mass part of the motion arises because the oscillator Hamiltonian has the property of separating into a part involving only the center of mass coordinate and a part involving only the relative coordinates, when center of mass coordinates are introduced. This is not true for other Hamiltonians with a space fixed potential. It is an important practical reason for choosing an oscillator potential instead of some other form, such as a square well, as a basis for shell model calculations, see for instance J. Elliot and T. Skyrme [9].

★★ For practical purposes we restore $Z\,(\boldsymbol{R}_{\mathrm{CM}})$ to $\psi_n$ because the $\psi_m$ have 8 particles and 24 degrees of freedom. Here $Z\,(\boldsymbol{R}_{\mathrm{CM}})$ will be an oscillator function with no excitation.

But in the single particle picture, if we are not to violate the Pauli principle, the lowest energy states of Be$^8$ must have four nucleons in the $1p$ shell. Then eq. (III,12a) shows that the oscillator cluster function must describe a state of at least four oscillator quanta of energy, i.e. $4\hbar\omega$ of energy above the zero point, because otherwise the cluster wave function would vanish under antisymmetrization. Since the $\alpha$-cluster parts of the wave function have been assumed to be without internal excitation, this means that in the oscillator cluster functions of lowest energies, the relative motion parts $\chi_k(\mathbf{R})$ are oscillator functions with four quanta of excitation energy and therfore must have orbital angular momentum quantum numbers of either $l = 0,2$ or $4$.

The next higher states of this type have six quanta of excitation in the relative motion part. The wave functions with five quanta of excitation in the relative motion parts vanish under antisymmetrization because the $\alpha$-clusters obey Bose statistics.* Other states of higher excitation are produced by internally exciting the $\alpha$-clusters.

We want to point out here that the method we have just used, of comparing nuclear wave functions in different expansion systems to discover antisymmetrization effects, is of quite general utility, and we shall use it frequently.

We have now used the oscillator potential to generate oscillator cluster trial wave functions. We are ready to consider these ansatz oscillator cluster wave functions in conjunction with the actual nuclear Hamiltonian, and to justify quantitatively our physical assumption — in the frame of the Ritz variational principle — that wave functions without internally excited $\alpha$-clusters are energetically preferred. In the nuclear Hamiltonian

$$H = \sum_{i=1}^{8} \frac{\mathbf{P}_i^2}{2M} + \sum_{i=1}^{8} \sum_{j>i}^{8} V_{ij} \tag{III, 13}$$

we choose for the interaction potential $V_{ij}$ a two-nucleon potential which fits in good approximation the nucleon-nucleon scattering data up to 100 MeV and gives also to a good approximation the correct deuteron binding energy, LEDERER [11], specifically:

$$V_{ij} = V_0\, e^{-\beta\,(\mathbf{r}_i - \mathbf{r}_j)^2} \cdot \left\{ w\left[1 - \frac{1}{4}(1+ \sigma_i \cdot \sigma_j)(1 + \tau_i \cdot \tau_j)\right] \right.$$
$$\left. + b\left[\frac{1}{2}(1 + \sigma_i \cdot \sigma_j) - \frac{1}{2}(1 + \tau_i \cdot \tau_j)\right]\right\} \tag{III, 14}$$

where

$$V_0 = -\,68.6\ \text{MeV};\ \ \beta = 4.17 \cdot 10^{25}\ \text{cm}^{-2};\ \ w = 0.41;\ \ b = 0.09$$

---

* This can be seen easily: If one exchanges the four particles of the one $\alpha$-cluster with the four particles of the other $\alpha$-cluster in every term of the antisymmetrized wave function, then due to the exchange of an even number of particles the relative motion parts of the wave function must not change their signs. On the other hand if $l$ is odd, which is connected with odd quanta of excitation in the relative motion parts, these parts will change their signs; the odd $l$ wave functions must vanish when antisymmetrized.

and the $\sigma$ and $\tau$ are the usual spin and isobaric spin operators which are expressed by the Pauli spin matrices $\sigma_x$, $\sigma_y$, $\sigma_z$.

We wish to make two remarks about the two-nucleon potential ansatz (III,14). First we have neglected for the moment spin-orbit and tensor forces. The tensor force while important for such sensitive effects as quadrupole moments, does not usually play a significant role in energetical considerations, central forces are found to be adequate in good approximation. The spin-orbit force does not usually make large contributions to the energy, and can be treated very often as a perturbation. In many of our examples we consider $\alpha$-clusters with no internal excitation, so that spin-orbit forces then strictly vanish. Hence we are justified as a first approximation in neglecting such forces.

A second more serious objection to the phenomenological potential (III,14) is that it does not have the saturation property. Hence in the variational calculation we cannot vary the oscillator width parameter, $a = M\omega/\hbar$. If this width parameter were actually varied, we would obtain collapsed states of small radius ($\approx 1.5 \cdot 10^{-13}$ cm), see for instance W. HEISENBERG [12]. Because we are making here a first rough calculation and wish to circumvent calculations with the analytically difficult hard-core forces, we will take into account the principle effects of the hard-core forces on the wave function (i.e. that it does not collapse) in an approximate way, as follows:

Calculations, especially in the many-body theory, K. BRUECKNER [13], L. GOMES et al. [14], A DE SHALIT and V. WEISSKOPF [15], W. BRENIG [16], indicate that two nucleon forces which become repulsive at short distances ($\approx 0.4 \cdot 10^{-13}$ cm) serve to prevent the collapse of nuclei, and so have a large effect on the nucleon density. However, they give approximately the same energy expectation values as obtained by fixing the nuclear radii phenomenologically and using forces with no repulsive core, F. SCHLOEGL [17], W. WILD and K. WILDERMUTH [18]. Hence we expect to obtain nearly correct energy results with the potential (III,14) so long as the oscillator width parameter $a$ is fixed to give the correct nuclear radius. There is a value of the oscillator width parameter ($a = 4.7 \cdot 10^{25}$ cm$^{-2}$) which gives nearly the right radius for all nuclei from He⁴ to O¹⁶, the increasing nuclear size being accounted for by the larger radii of the higher oscillator levels, B. CARLSON and I. TALMI [19]. Fixing the oscillator width $a$ to this value is a preliminary phenomenological way giving reasonable results to describe the saturation character of the nuclear forces.

What the fixing of the nuclear radii in our variational calculations means can also be considered from a slightly different point of view which gives further insight into the limitations of these preliminary calculations. If we include in the variational calculations with fixed oscillator width parameter ($a \approx 4.7 \cdot 10^{25}$ cm$^{-2}$) all possible excited oscillator cluster states, then we still obtain for the energetically lowest states collapsed states, because our oscillator cluster wave functions form a complete set of functions. The amplitudes of the excited oscillator states will be small, but they converge so slowly with increasing values of the energy that the main parts of the wave functions of the low excited nuclear states will be formed by a superposition of highly excited oscillator cluster wave functions. The superposition will be such that for large $r$-values the cluster wave functions will cancel each other by interference and only for $r \leq 1.5 \cdot 10^{-13}$ cm will the nuclear wave functions for the low excited states be appreciably different from zero. Certainly the kinetic energy of such collapsed states will be relatively large due to the small radii of the collapsed states* (uncertainty principle), but this energy is overcompensated by the large negative potential energy, which comes from the concentration of all nucleons in a small volume of the dimension of their mutual interaction range thereby bringing the potentials for the strongly attractive two-nucleon forces to a minimum. Therefore to avoid the collapse of our nuclear states without introducing forces with repulsive core we have not only to fix the width parameter of the oscillator potential but also we have to restrict (at this stage of our considerations) our oscillator cluster wave functions used in the variational calculations to relatively low excited cluster states.

To ignore the repulsive core of the nuclear forces is certainly very undesirable from a theoretical point of view. But if we include this repulsive core in our potential, then our wave functions must either vanish or become very small whenever two nucleons approach each other within a separation less than the repulsive core radius, which means that the simple type of oscillator trial wave functions used above will not work. Later we will generalize our cluster wave functions such that the influence of the repulsive core of the two-nucleon forces can be included in our considerations in a consistent manner. This generalization can be done in a very natural way.

We consider now the variational calculation of the low energy states of Be$^8$ with the nuclear Hamiltonian (III,13–14). Since we have fixed the oscillator width parameter and since waves of different angular momenta cannot be mixed, there are no degrees of freedom left in our cluster functions and no variation is actually carried out in this example. The cluster wave functions (III,15) are our approximate eigenfunctions; it only remains to inquire what their energies are.

Computing expectation values of the Hamiltonian (III,13–14) we first find that the states with six or more oscillator excitation quanta in the relative motion, or with internally excited $\alpha$-clusters, have energy expectation values at least 20 MeV higher than states with four excitation quanta in the relative motion and no internally excited $\alpha$-clusters. Furthermore, the overlap integrals, $\langle s|H|j \rangle$, (see footnote following eq. (II,7)) of the lowest states with the higher states are relatively small when compared to the energy gap $E_s - E_j$ between these states. For these reasons we will retain in the Ritz variational calculation only the latter kind of states. The amplitudes of the high-energy cluster waves are expected to be small. With this we have justified our assumption

---

* In the variational calculation this kinetic energy comes from the high excited oscillator cluster states.

about the high internal binding energies of the $\alpha$-clusters. Our cluster wave function is now reduced to

$$\psi_n(\boldsymbol{R}, \overline{\boldsymbol{r}}_1, \ldots \overline{\boldsymbol{r}}_8)$$
$$= \mathscr{A} \; \phi_0(\overline{\boldsymbol{r}}_1, \ldots \overline{\boldsymbol{r}}_4) \; \phi_0(\overline{\boldsymbol{r}}_5, \ldots \overline{\boldsymbol{r}}_8) \; \chi_{4\hbar\omega, l, m_l}(\boldsymbol{R}) \; \xi(s_1, \ldots t_8) \qquad \text{(III, 15)}$$

and is seen to be 15-fold degenerate with respect to the oscillator Hamiltonian, with the following $l$ and $m_l$ values:

$$l = 0, \; m_l = 0 \; ;$$
$$l = 2, \; m_l = -2, -1, 0, 1, 2; \qquad \text{(III,16)}$$
$$l = 4, \; m_l = -4, \ldots, 4 \; .$$

The kinetic energy operator of the nuclear Hamiltonian is the same as that in the oscillator Hamiltonian when the total center of mass kinetic energy of the nucleons in the oscillator well is split off; the contribution of the kinetic energy to the energy expectation value is the same for all angular momenta. Calculating the expectation values of the potential energy operator given above, we obtain total energy expectation values such that $(E_{l=4} - E_{l=0}) = 12.5 \text{ MeV}$ and $(E_{l=2} - E_{l=0}) = 3.75 \text{ MeV}$ [8]. These values compared with the experimental values are 10–20% too large, which is a satisfactory agreement for a calculation of this sort. The ordering of levels can be understood qualitatively, for with our attractive short-range forces the energy will be lowest for the states with the most overlap of the $\alpha$-clusters, that is, for states of lowest angular momentum (penetrating orbit argument).

This completes our calculations on the low lying states of Be⁸, which we have used to illustrate the method of cluster representations. To summarize, first we used our physical knowledge to decide the best cluster representation for these states. With this representation we used the oscillator potential to generate the appropriate expansion system of oscillator cluster functions for our trial wave functions. Finally, we verified our assumptions quantitatively by showing that in the trial nuclear wave function, which gave reasonable energy expectation values with the aid of the Ritz principle, terms with broken-up $\alpha$-clusters or higher motion excitation, had only small amplitudes.

In our considerations of Be⁸, as mentioned earlier, we have neglected the spin-orbit coupling term in the nuclear forces. For the low excited states of Be⁸ this is allowed to a very good approximation; in the unexcited $\alpha$-cluster wave functions which we have used here, the spin-orbit coupling can give no direct contribution to the expectation value of the energy because of the spherical symmetry of these $\alpha$-clusters. The influence of the spin-orbit coupling on the energy expectation value of these states therefore can only be indirect, by mixing some broken-up $\alpha$-cluster states into the cluster states with ground state $\alpha$-clusters and low excitation in the relative motion. The contribution of such broken-up states to the low excited states of Be⁸ is very small due to the tight binding of the unexcited $\alpha$-clusters.

We now want to point out some surprising features in our Be⁸ calculations which arise from antisymmetrization. We have seen that the shell model and the cluster method arise from very different physical viewpoints. Yet due to the effect of antisymmetrization our cluster calculations for the three lowest Be[8] levels are completely equivalent to the corresponding shell model calculations, D. KURATH [20], if one neglects there the spin-orbit coupling terms in the Hamiltonian [8].

Let us point out another surprising agreement between apparently different models, K. WILDERMUTH and TH. KANELLOPOULOS [21]. We see in our cluster calculation of Be[8] that the ratio $(E_{l=4} - E_{l=0})/$ $(E_{l=2} - E_{l=0})$ is exactly 10/3. This ratio is independent of the choice of $V_0, w, b,$ and $\beta$ used in the two-nucleon potential (III,14), E. FEENBERG and M. PHILLIPS [22], G. RACAH [23], provided only that the wave functions describing these states are pure oscillator cluster wave functions of the form defined above, see eq. (III,3)−(III,12). One obtains the same ratio for these three energy states if one describes Be[8] by an $\alpha$-particle dumb-bell model and considers these states as the three lowest rotational states of the Be[8] dumb-bell. In the oscillator-cluster viewpoint the energy differences of the states are entirely due to potential energy, but in the Be[8] dumb-bell model they are entirely due to kinetic energy. The reason for this paradoxical agreement in the energy levels of these two different Be[8] models also comes from the antisymmetrization of the wave functions, which under specific circumstances makes these two models completely equal to each other, as we shall see later.

We turn now to a first consideration of the effects of antisymmetrization on the cluster wave functions.


### D. Effects of antisymmetrization

We have already mentioned that as a result of antisymmetrization seemingly quite different wave functions can become very similar or even equivalent to each other. This is an important key to the resolution of the apparently contradictory description of nuclei by different models, all of which have had some success in predicting nuclear characteristics. Here we will illustrate the effects of antisymmetrization on two specific examples.

As first example let us consider a large number of fermions without mutual interaction in their ground state in a square well potential [8]; this would approximately describe the electrons of a conductor for example. The energy differences among the particles in the well arise from their kinetic energies, because the energies of the states increase with the square of their momenta. In the ground state all single particle states are filled inside a momentum sphere* known as the Fermi sphere (see Fig. 1-A). If this system as a whole is now given a small velocity $\Delta v$, the Fermi sphere is slightly shifted so that its center is no longer at the origin (see Fig. 1-B). The change relative to the situation in Fig. 1-A is a collective excitation in which each fermion receives a small change in momentum, $m \cdot \Delta v$. Now let us instead start with the

---

* We use the well known approximation of plane wave states.

Fermi sphere at the origin (as in Fig. 1-A) and impart various large amounts of momentum to a few of the fermions (all those in states in region 1 of Fig. 1-C) at the left of the sphere so as to excite them into states just to the right of the sphere (filling the states in region 3 of Fig. 1-C). Due to the indistinguishability of the fermions, corresponding to the antisymmetrization of the wave function, the situation in Fig. 1-C is completely equivalent to that in Fig. 1-B. This shows how under antisymmetrization a large excitation imparted to a few fermions can be equivalent to a collective excitation of all the fermions as a whole.

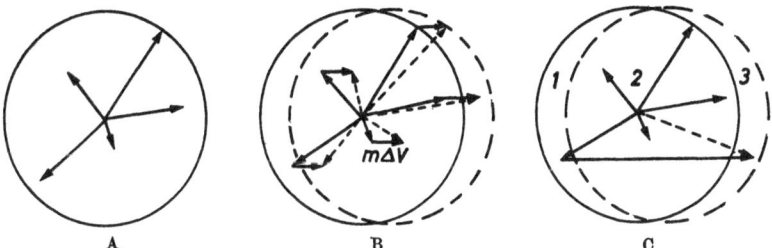

Fig. 1 A—C. Example of equivalence under antisymmetrization. A. The arrows denote representative states of the filled Fermi sphere. B. All particles shifted by amount $\Delta p = m \, \Delta v$. C. Just those particles in region 1 shifted to region 3

Quantitatively the equivalence of these two pictures comes from the fact that when the collective wave function of the situation 1-B is antisymmetrized many terms will cancel each other so that the final antisymmetrized wave function becomes equal to the antisymmetrized wave function of situation 1-C*. This is, for instance, the reason why a

---

* Quantitatively, consider $N$ fermions in a well with periodic boundary conditions and level spacing $\Delta k$. Then the ground state wave function is

$$\psi = \mathscr{A} \exp\left( \sum_{n=-f}^{+f} i\, n \, \Delta k \, x_n \right); \quad 2f + 1 = N$$

where for simplicity and clarity we consider only one-dimension. $\mathscr{A}$ is the antisymmetrizing operator. In one dimension the Fermi sphere reduces to a line of points in momentum space, centered about the origin and filled up to some number $\pm f$.

Now let us give each fermion an additional momentum of $\Delta p = \hbar \, \Delta k$, so that the total momentum added to the system is $N \, \Delta p$. The new wave function becomes

$$\psi = \mathscr{A} \exp\left( \sum_{n=-f}^{f} i \, (n \, \Delta k + \Delta k) \, x_n \right).$$

The center of the "Fermi-line" is shifted $\Delta p$ to the right of the origin. But we can write the new function as

$$\psi = \mathscr{A} \exp\left( \sum_{n=(-f+1)}^{f+1} i \, n \, \Delta k \, x_n \right).$$

Because the labeling of the particles is irrelevant under antisymmetrization, this is the same as giving the left most particle all the momentum $N \, \Delta p$ and exciting it from the state $\exp\{-i\,f\,\Delta k\,x$ to the state $\exp\{i\,(f+1)\,\Delta k\,x\}$ to the right of the original "Fermi-line". Thus what we originally conceive as a collective excitation of all the fermions moving together to the right, and as a single particle excitation of one fermion particle strongly excited to the right, become equivalent under antisymmetrization.

2*

collective dipole oscillation can be exactly equivalent to a one-particle excitation, D. BRINK [10].

As second example, we will explicitly carry out the antisymmetriza-tion in the relatively simple case of the $l = 4$, $l_z = 4$, $\alpha$-cluster state of Be⁸ discussed earlier and show how the wave function looks when ex-pressed as a superposition of single particle shell model wave functions, TH. KANELLOPOULOS and K. WILDERMUTH [24, 25].

The cluster wave function for this state, which we shall write out in detail below, was taken to be

$$\psi = \mathscr{A}\,\phi_0(\underset{\sim}{\alpha_1})\,\phi_0(\underset{\sim}{\alpha_2})\,\chi_{4,4,4}(R)\,Z_0(R_{\mathrm{CM}})\,. \tag{III,17}$$

Here we include the physically irrelevant center of mass motion of Be⁸, $Z_0(R_{\mathrm{CM}})$ in our calculations in order to make the transition to the single particle shell model wave functions.

The $\alpha$-cluster internal functions are products of single-particle wave functions of the relative internal coordinates $\bar{r}_i$ and the spin and isobaric spin coordinates:

$$\phi_0(\underset{\sim}{\alpha}) = \prod_{i=1}^{4} \phi_{\mathrm{space}}^i(\bar{r}_i)\cdot\phi_{\mathrm{spin}}^i(s_i)\cdot\phi_{\mathrm{isospin}}^i(t_i)\,. \tag{III,18}$$

Because in Be⁸ the $\alpha$-clusters have no internal excitation, $\phi_{\mathrm{space}}^i(\bar{r}_i)$ correspond to a $1s$ oscillator wave function, $\mathrm{Const.}\cdot\exp\!\left(-\dfrac{a}{2}\,\bar{r}^2\right)$, see App. A, eq. (A,7). The functions $\phi_{\mathrm{spin}}^i(s_i)$ and $\phi_{\mathrm{isospin}}^i(t_i)$ correspond to two protons and two neutrons with spins up and down in such a way as not to violate the Pauli principle:

$$\prod_{i=1}^{4}\phi_{\mathrm{spin}}^i\phi_{\mathrm{isospin}}^i = \binom{1}{0}_{s_1}\binom{1}{0}_{t_1}\binom{0}{1}_{s_2}\binom{1}{0}_{t_2}\binom{1}{0}_{s_3}\binom{0}{1}_{t_3}\binom{0}{1}_{s_4}\binom{0}{1}_{t_4}\,. \tag{III,19}$$

As discussed earlier, the relative motion function must have at least four quanta of excitation. For definiteness and simplicity, we will take $\chi(R)$ as a $1g$ oscillator wave function (see App. B) of the relative coordinate $R$ of the two clusters, with $l = 4$ and $m_l = 4$:

$$\chi(R) = \mathrm{const.}\cdot e^{-aR^2}R^4\,Y_{44}(\theta_R,\,\varphi_R) \tag{III,20}$$

where $Y_{44}(\theta_R,\,\varphi_R)$ is the spherical harmonic with the indices $l = 4$, $m_l = 4$ and $(R,\,\theta_R,\,\varphi_R)$ are the polar coordinates of $R$. Finally $Z_0(R_{\mathrm{CM}})$ corresponds to a zero oscillation of the center of mass of Be⁸:

$$Z_0(R_{\mathrm{CM}}) = e^{-4aR_{\mathrm{CM}}^2}\,. \tag{III,21}$$

The exponentials in $\chi(R)$ and $Z(R_{\mathrm{CM}})$ result by solving the separated oscillator Schrödinger equations (see eq. III,7), obtained by introducing the different relative coordinates and center of mass coordinates of the clusters in Be⁸. With (III,18)−(III,21) we get the following form for the cluster wave function $\psi$ of the $Be^8$ state considered here:

$$\psi = \mathscr{A}\left\{\left[\exp\left\{-\frac{a}{2}\sum_{i=1}^{4}\bar{r}_i^2\right\}\binom{1}{0}_{s_1}\binom{1}{0}_{t_1}\binom{0}{1}_{s_2}\binom{1}{0}_{t_2}\binom{1}{0}_{s_3}\binom{0}{1}_{t_3}\binom{0}{1}_{s_4}\binom{0}{1}_{t_4}\right]\times\right.$$

$$\times\left[\exp\left\{-\frac{a}{2}\sum_{i=5}^{8}\bar{r}_i^2\right\}\binom{1}{0}_{s_5}\binom{1}{0}_{t_5}\binom{0}{1}_{s_6}\binom{1}{0}_{t_6}\binom{1}{0}_{s_7}\binom{0}{1}_{t_7}\binom{0}{1}_{s_8}\binom{0}{1}_{t_8}\right]\times \quad (III,22)$$

$$\left.\times\,[e^{-aR^2}R^4\,Y_{44}(\theta_R,\,\varphi_R)]\cdot[e^{-4a\,R^2_{CM}}]\right\}.$$

$\mathscr{A}$ indicates antisymmetrization and normalization.

Let us now express the cluster wave function in terms of particle coordinates $r_i$ so that we may carry out the antisymmetrization explicitly.

If we express $\boldsymbol{R}$, $\boldsymbol{R}_{CM}$, $\bar{r}_i$ in terms of the original coordinates $r_i$ by means of formulas (III,5), (III,6) and (III,10) then the exponents simplify:

$$\left\{\frac{a}{2}\sum_{i=1}^{4}\bar{r}_i^2+\frac{a}{2}\sum_{i=5}^{8}\bar{r}_i^2+a\,R^2+4\,a\,R^2_{CM}\right\}$$

$$=\frac{a}{2}\left\{\sum_{i=1}^{4}(r_i-R_1)^2+\sum_{i=5}^{8}(r_i-R_2)^2+2(R_1-R_2)^2+2(R_1+R_2)^2\right\}$$

$$=\frac{a}{2}\left\{\sum_{i=1}^{8}r_i^2-2R_1\cdot\sum_{i=1}^{4}r_i-2R_2\cdot\sum_{i=5}^{8}r_i+8R_1^2+8R_2^2\right\} \quad (III,23)$$

$$=\frac{a}{2}\sum_{i=1}^{8}r_i^2.$$

$e^{-\frac{a}{2}\sum_{i=1}^{8}r_i^2}$ is a totally symmetric function of the eight coordinates and therefore can be ignored while carrying out the antisymmetrization. Note that the simple form of the exponential factor is a consequence of having retained the zero-th order oscillation of the center of mass. We are left now with

$$\mathscr{A}\left\{R^4\,Y_{4,4}(\theta_R,\,\varphi_R)\cdot\binom{1}{0}_{s_1}\cdots\binom{0}{1}_{t_8}\right\}$$

$$=\mathscr{A}\left\{R^4\sin^4\theta_R(\cos\varphi_R+i\sin\varphi_R)^4\cdot\binom{1}{0}_{s_1}\cdots\binom{0}{1}_{t_8}\right\}$$

$$=\mathscr{A}\left\{(X+iY)^4\cdot\binom{1}{0}_{s_1}\cdots\binom{0}{1}_{t_8}\right\}$$

$$=\mathscr{A}\left\{\left(\frac{1}{4}\right)^4[(x_1+iy_1)+(x_2+iy_2)+\cdots-(x_5+iy_5)\cdots-(x_8+iy_8)]^4\times\right.$$

$$\left.\times\binom{1}{0}_{s_1}\cdots\binom{0}{1}_{t_8}\right\}=\mathscr{A}\left\{\left[\text{a sum of terms of the form:}\right.\right.$$

$$\text{const}\cdot(x_j+iy_j)^{n_j}\cdot(x_k+iy_k)^{n_k}\ldots(x_l+iy_l)^{n_l}\right]\cdot\binom{1}{0}_{s_1}\cdots\binom{0}{1}_{t_8}\right\} \quad (III,24)$$

where $j,k,l=1,2,\ldots 8$; and $n_j+n_k+\cdots+n_l=4$.

However, when antisymmetrization is performed, terms with $n_i$ different from 1 disappear. In agreement with the Pauli principle, only terms corresponding to four nucleons in the $1p$ shell (whose wave functions besides their exponential factors are proportional to $x_1 + iy_1$) and four nucleons in the $1s$ shell (whose wave functions, besides their exponential factors are constants) are different from zero. Thus for example, if we have the term

$$(x_1 + iy_1)^2 \, (x_2 + iy_2) \, (x_3 + iy_3) \begin{pmatrix} 1 \\ 0 \end{pmatrix}_{s_1} \cdots \begin{pmatrix} 0 \\ 1 \end{pmatrix}_{t_8}$$

and we interchange particles 4 and 8, we obtain another term equal to the original one but with the opposite sign, as the corresponding permutation is odd. Then these terms cancel each other.

The antisymmetrized wave function can thus be written:

$$\mathscr{A}\,\psi = \mathscr{A}\,N \left\{ (x_5 + iy_5) \, (x_6 + iy_6) \, (x_7 + iy_7) \, (x_8 + iy_8) \times \right.$$
$$\left. \times \, e^{-\frac{a}{2} \sum\limits_{i=1}^{8} r_i^2} \begin{pmatrix} 1 \\ 0 \end{pmatrix}_{s_1} \cdots \begin{pmatrix} 0 \\ 1 \end{pmatrix}_{t_8} \right\} \tag{III,25}$$

with $N$ a normalization factor. We see that this wave function corresponds to four $1p$ nucleons with parallel orbital angular momenta and four $1s$ nucleons. And this is the way it had to be, since we started with a total orbital angular momentum $l = 4$.

We have thus shown that the antisymmetrized cluster function (III,17) is completely identical to the antisymmetrized shell model function (III,25). This gives us a first clue to the resolution of apparently contradictory physical descriptions of the same nuclear state. We shall return to this point again and again.

We now want to give a frequently useful method for finding the effect of antisymmetrization on the total spin $S^2$ and isobaric spin $T^2$ of the nucleons. We will introduce this method using our $Be^8$ state (III,25) as a specific example.

We will show that the antisymmetrized wave function (III,25) has the eigenvalues $S^2 = 0$ and $T^2 = 0$ respectively, notwithstanding that the unsymmetrized wave function (III,25) is an eigenfunction of $S_z$ and $T_z$ with eigenvalues $S_z = 0$, $T_z = 0$ but not an eigenfunction of $S^2$ or $T^2$. This is easily proved using the following property of the antisymmetrizer operator $\mathscr{A}$, where $P_0$ is any fixed permutation:

$$\mathscr{A}\,P_0 = [\sum_P (-1)^P P] \, P_0$$
$$= (-1)^{P_0} \sum_P (-1)^{P\,P_0} P\,P_0 = (-1)^{P_0}\mathscr{A}\,. \tag{III, 26}$$

The formula (III,26) follows from the fact that $P \cdot P_0$ ranges over all permutations when $P$ ranges over all permutations.

This permits us to write:

$$\mathscr{A}\,\psi = \text{const} \cdot \mathscr{A}\,(\psi + C \cdot P_0\,\psi) \tag{III,27}$$

provided only that we do not choose the constant $C$ so as to make the right side vanish identically. Now the spin and isobaric spin part of the function $\psi$ upon which the permutations act in (III,25) may be written out in full as

$$\begin{pmatrix}1\\0\end{pmatrix}_{s_1}\begin{pmatrix}1\\0\end{pmatrix}_{t_1}\begin{pmatrix}0\\1\end{pmatrix}_{s_2}\begin{pmatrix}1\\0\end{pmatrix}_{t_2}\begin{pmatrix}1\\1\end{pmatrix}_{s_3}\begin{pmatrix}0\\1\end{pmatrix}_{t_3}\begin{pmatrix}0\\1\end{pmatrix}_{s_4}\begin{pmatrix}0\\1\end{pmatrix}_{t_4}\times$$

$$\times\begin{pmatrix}1\\0\end{pmatrix}_{s_5}\begin{pmatrix}1\\0\end{pmatrix}_{t_5}\begin{pmatrix}0\\1\end{pmatrix}_{s_6}\begin{pmatrix}1\\0\end{pmatrix}_{t_6}\begin{pmatrix}1\\0\end{pmatrix}_{s_7}\begin{pmatrix}0\\1\end{pmatrix}_{t_7}\begin{pmatrix}1\\1\end{pmatrix}_{s_8}\begin{pmatrix}0\\1\end{pmatrix}_{t_8}$$

(III, 28)

and we note that the space part is unchanged by permutations either among the first four nucleons or among the last four nucleons. Now choosing $P_0$ as $P_{1,3}$ the permutation of nucleons $1$ and $3$, and $C$ as $-1$, we obtain

$$\mathscr{A}\,\psi = \text{const}\cdot\mathscr{A}\left\{\left[\begin{pmatrix}1\\0\end{pmatrix}_{t_1}\begin{pmatrix}0\\1\end{pmatrix}_{t_3}-\begin{pmatrix}0\\1\end{pmatrix}_{t_1}\begin{pmatrix}1\\0\end{pmatrix}_{t_3}\right]\times\right.$$

$$\left.\times\begin{pmatrix}1\\0\end{pmatrix}_{s_1}\begin{pmatrix}1\\0\end{pmatrix}_{s_2}\begin{pmatrix}0\\1\end{pmatrix}_{s_3}\begin{pmatrix}1\\0\end{pmatrix}_{t_2}\cdots\begin{pmatrix}0\\1\end{pmatrix}_{s_8}\begin{pmatrix}0\\1\end{pmatrix}_{t_8}\right\}$$

(III, 29)

which corresponds to a coupling of nucleons $1$ and $3$ into an eigenfunction of total isobaric spin zero. Proceeding further, we may write

$$\mathscr{A}\,\psi = \text{const}\ \mathscr{A}\left\{(1-P_{6,8})\,(1-P_{5,7})\,(1-P_{2,4})\,(1-P_{1,3})\,\psi\right\}\qquad\text{(III,30)}$$

in which the right-side bracket has nucleon pairs $(1,3)$, $(2,4)$, $(5,7)$, and $(6,8)$ all coupled into an eigenfunction of isobaric spin-zero. This shows[*] that the wave function $\mathscr{A}\,\psi$ must be a $T^2 = 0$ eigenstate. Now one can start again and couple pairs $(1,2)$, $(3,4)$, $(5,6)$ and $(7,8)$ into spin zero eigenfunctions, thus showing that $\mathscr{A}\,\psi$ is also an eigenstate of $S^2 = 0$. This type of argument is useful in many problems because very often one can start with unsymmetrized wave functions which are not eigenfunctions of total spin and isobaric spin but become eigenfunctions automatically after antisymmetrization. Note that the behaviour of the space part of the wave function is very important. If the space part would have changed its sign in the permutation $P_{1,3}$ then this permutation would have created a $T = 1$ state. If the space part of the wave function is not an eigenfunction of such permutations, then the method does not work and we have to start from the beginning with eigenfunctions of the spin and isobaric spin operator.

Let us now summarize these first remarks on the effects of the Pauli principle. Antisymmetrization of the nuclear wave function is the quantitative expression of the fact that nucleons are indistinguishable fermions. We must beware of too naive conceptions of single particle and collective motions; such conceptions often arise from a too literal interpretation of unsymmetrized wave functions. Only what remains of

---

[*] Since $T^2 = (\boldsymbol{T}_1 + \boldsymbol{T}_2 + \cdots + \boldsymbol{T}_8)^2$ is symmetric in all particle coordinates, it commutes with $P_{i,j}$ and hence with $\mathscr{A}$. If for arbitrary $\psi$, $T^2\,\psi = t^2\,\psi$ with eigenvalue $t^2$ then

$$T^2\,\mathscr{A}\,\psi = \mathscr{A}\,T^2\,\psi = t^2\,\mathscr{A}\,\psi\,,$$

that is, if $\psi$ is an eigenfunction to $T^2$ with eigenvalue $t^2$, so is $\mathscr{A}\,\psi$.

such conceptions after antisymmetrization constitutes the actual behaviour of the system and may not resemble our original ideas at all. Indeed
as we have seen, very different starting points can lead upon antisymmetrizing to identical or nearly identical wave functions; such effects
help us to resolve the apparent differences among the many nuclear
models. The generalized cluster method does help to build a picture
whose more general features will survive antisymmetrization but we must
be cautious in taking any picture too literally. The oscillator cluster
model, which is the theory we have developed so far, has the advantage
of providing several different viewpoints, mathematically equivalent
after antisymmetrization, from which to regard each nuclear phenomenon.
By extracting the noncontradictory features from the various viewpoints
(see for example text following eq. (III,12a)) we are guided to the correct
physical behaviour. Further effects of antisymmetrization, and especially
the physical consequences, will be discussed frequently in the following
pages.

## E. The application of the oscillator cluster model to the energy spectra of other light nuclei

In this chapter we will discuss qualitatively the low lying energy
levels of some other light nuclei from the point of view of the oscillator
cluster model. We shall obtain trial wave functions consisting of one or
two oscillator cluster functions which correctly predict the spins, parities
and, used in conjunction with the phenomenological potential, the qualitative ordering of the nuclear levels. Although we cannot expect particularly good quantitative agreement with experiment from such simple
trial functions, the qualitative success of so simple a model can give
considerable insight into nuclear phenomena. Moreover from the discussion we will obtain indications on how to generalize the oscillator cluster
functions so as to later obtain quantitative agreement as well.

As first example we consider the groundstate and the low excited
states of $Li^7$ and $Be^7$.

From our general considerations above the most appropriate wave
functions for the description of the low $Li^7$ states are the $\alpha$-cluster *triton*-
cluster wave functions, D. INGLIS [26, 24], G. PHILLIPS and T. TOM-
BRELLO [27], H. HOLMGREEN and L. CAMERON [28], Y. TANG et al. [29],
which have the form

$$\psi_{ijk} = \mathscr{A}\, \phi_i(\underset{\sim}{\alpha})\; \phi_j(\underset{\sim}{t})\; \chi_k(\alpha - t)$$

or                                                                                          (III,31)

$$\psi = \sum_{ijk} a_{ijk}\, \psi_{ijk}$$

where $\phi_i(\underset{\sim}{\alpha})$ and $\phi_j(\underset{\sim}{t})$ are the internal wave functions of the *alpha* and
*triton* clusters (including the spin and isobaric spin coordinates of the

clusters) and $\chi_k(a - t)$ is the relative motion function of the clusters. The separation energy of the last nucleon in a free triton is about 6 MeV, which means that the internal binding energy of a triton is rather large. Therefore we can presume Li⁷ states which contain internally excited *triton* clusters with an appreciable amplitude should start in the region of 5 MeV excitation energy, which means we can assume $i = 0$ and $j = 0$ for the low excited states. We should also mention that the overlap integrals $\langle s \mid H \mid i \rangle$ between oscillator cluster functions with unexcited clusters and those with a broken up *triton* cluster are relatively small compared with this 5 MeV energy gap. In the oscillator cluster model the relative motion function $\chi(a - t)$ of lowest energy which does not vanish under antisymmetrization will correspond to three oscillator quanta of excitation, i.e. it will be an oscillator function of energy★ $\left(3 + \dfrac{1}{2}\right)\hbar\omega$. This follows since the wave function in the single-particle oscillator picture has three $1p$ nucleons or an energy of three quanta, whereas the *triton* and *alpha* clusters in their center of mass frames contain only $1s$ nucleons and so have zero quanta of internal energy; the relative motion must therefore contain three quanta, since the cluster wave function has then the same oscillator energy as the corresponding single-particle oscillator wave function.

Oscillations with three quanta can be either $2p$ or $1f$ oscillations (see Appendix B) with orbital angular momenta $l = 1$ and $l = 3$ respectively. By the same "penetrating orbit" arguments used in the case of Be⁸, we expect the states with $l = 1$ to be lower in energy than the states with $l = 3$. In actuality these states are further split by the spin-orbit coupling.

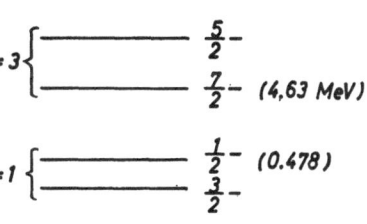

Fig. 2. Level scheme for Li⁷

The triton spin being $\dfrac{1}{2}$, the total angular momentum for the $l = 1$ states can be $\dfrac{1}{2}$ or $\dfrac{3}{2}$. The spin-orbit coupling demands that the lowest state must correspond to parallel spin and orbital angular momenta, that is, the lowest state has $I = \dfrac{3}{2}$. The order of the $l = 3$ states with the sequence $I = \dfrac{7}{2}$, $I = \dfrac{5}{2}$ is similarly established. The experimental sequence of the low lying energy levels of Li⁷, see F. AIZENBERG-SELOVE and T. LAURITSEN [30] and Fig. 2, is in agreement with this qualitative

---

★ We emphasize this is only an oscillator energy, obtained by operating on the oscillator cluster function with the oscillator Hamiltonian. The desired approximate energy of the corresponding nuclear level is obtained by using this oscillator cluster function as a trial function in a variational calculation with the phenomenological nuclear Hamiltonian. Insofar as the oscillator potential represents the averaged nuclear potential, the oscillator energy is a rough (but only a rough) indication of the corresponding approximate nuclear energy.

description*. All these levels have negative parity due to the odd orbital angular momentum of the relative motion of the clusters and the even internal parity of the $\alpha$-cluster and $t$-cluster.

Be$^7$ possesses a low energy spectrum analogous to Li$^7$ only that all levels are lifted up about 1 MeV due to the larger Coulomb energy.

As second example we consider the Li$^6$ nucleus. This is a very interesting nucleus both from the theoretical and experimental point of view. It is a stable nucleus (as opposed to the lighter He$^5$ or Li$^5$) containing on the one hand sufficiently many nucleons to exhibit many important general features of nuclear phenomena, and on the other hand sufficiently few that detailed calculations can be carried out on it even for nuclear forces with a hard core. Therefore Li$^6$ is a test nucleus in nuclear physics much like the hydrogen atom has been in atomic physics. We shall return again and again to the Li$^6$ nucleus to test how refinements in our considerations improve our description of it.

We begin our description of the low energy states of Li$^6$ by assuming an $\alpha$-cluster *deuteron*-cluster structure, A. GALONSKY and M. McELLISTREM [31, 29], D. JACKSON and L. ELTON [32]. For the lowest energy levels we assume again that the $\alpha$-cluster has no internal excitation. But it is not immediately obvious that a $d$-cluster in a larger nucleus will not be essentially broken up, because the binding energy of a free deuteron in the triplet state ($S = 1$, $T = 0$) is only 2.2 MeV and the singlet state ($S = 0$, $T = 1$) exists only as a virtual state (resonance energy $\approx$ 64 KeV).

We can make the following plausible argument for the stability of a $d$-cluster in a larger nucleus, K. WILDERMUTH [33]. Because of the strong short range character of the nuclear forces, the potential energy of interaction of the two nucleons in a free deuteron is quite large ($\approx$ 15 MeV for the singlet state and $\approx$ 25 MeV in the triplet state). But due to the small volume of the deuteron, the uncertainty principle demands a correspondingly large internal kinetic energy of the two nucleons, so that the binding energy of the free deuteron is quite small. Now inside a nucleus the effect of the $d$-cluster structure is to introduce a correlation amongst the nucleons such that the two nucleons constituting the $d$-cluster, whichever they are at any moment, have a high probability of being close to one another**. Hence the $d$-cluster should retain the high internal potential and kinetic energies characteristic of the free deuteron.

---

* Calculations indicate the $\dfrac{5}{2}^-$ level (cluster structure $\alpha - t$) is rather broad ($\approx$ 1 MeV), see [29] for theoretical discussion and for experimental evidence see [168]. The relatively sharp $\dfrac{5}{2}^-$ level at 7.47 MeV has a different cluster structure (Li$^5 - n$) in which the triton cluster is broken up, see KHANNA et al. [169].

** In the oscillator cluster model this correlation is not particulary strong as yet. Later, when we generalize the cluster functions, we shall see that by strengthening the $d$-cluster correlation our description of Li$^6$ improves in the frame of the Ritz variational principle. We should also point out that, because the nucleons are indistinguishable, exactly which nucleons participate in the $d$-cluster motion can change from instant to instant.

If, however, the $d$-cluster is broken up, the potential energy between the two now uncorrelated nucleons drops drastically. But because the presence of the other cluster (in Li[6] the $\alpha$-cluster) confines the two uncorrelated nucleons within the nuclear volume, the uncertainty principle does not permit a correspondingly great decrease in their kinetic energy*. Hence due to this large kinetic energy we conclude that the Li[6] states with an unexcited $\alpha$-cluster and a broken-up $d$-cluster must lie several MeV above the states where neither cluster is broken up. A rough calculation with the Serber ansatz (III,14) shows this energy gap to be about 5 MeV, corroborating our arguments. As in Li[7] the overlap integrals $\langle s \mid H \mid j \rangle$ with these higher states are also again relatively small compared with this energy gap. Therefore in first approximation we can neglect the contributions of such broken-up $d$-cluster states to the low lying $T = 0$ and $T = 1$ states of Li[6]. We therefore can proceed assuming that the low excited Li[6] states in first approximation can be described by $\alpha$-cluster deuteron-cluster wave functions without internal excitation of these clusters. By considerations similar to those used in the cases of Be[8] and Li[7], we deduce that an oscillator energy of two quanta is associated with our cluster wave function ($2s$ or $1d$ oscillation, see Appendix B). Since the clusters have no internal excitation, this energy corresponds to the relative oscillation of the two clusters. The orbital angular momentum can be then $l = 0$ or $l = 2$. The first one corresponds to the ground state and the second one to three excited states associated with total angular momenta $I = 3, 2, 1$ resulting from the addition of the orbital angular momentum $l = 2$ and the deuteron cluster spin $S = 1$ in the triplet state ($T = 0$). Spin-orbit interactions as before, give the sequence shown in Fig. 3.

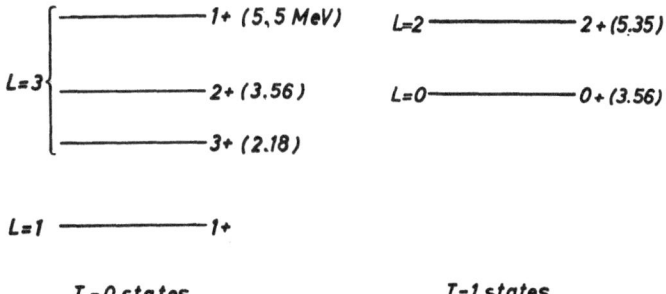

Fig. 3. Level scheme for Li[6]

The low-energy $T = 1$ states of Li[6] must be built of an unexcited $\alpha$-cluster and the deuteron cluster excited to the lowest $T = 1$, $S = 0$ state. In the single particle picture recoupling the spins to obtain a $T = 1$ state still leaves two nucleons in the $1p$ shell. Hence in the

---

* In the oscillator cluster model the total kinetic energy (internal kinetic energies of the $\alpha$- and $d$-clusters plus kinetic energy of the relative motion of the clusters) cannot change unless we permit excitations outside of the $1p$ shell of the single-particle oscillator states.

oscillator cluster picture the relative motion must correspond to two quanta of energy and therefore will have again an orbital angular momentum $l = 0$ or $l = 2$. This will also represent the total angular momenta of the states since both clusters have zero spin. The $I = 0$ state will lie lower according to penetrating orbit arguments; it will lie several MeV above the $I = 0$, $T = 0$ ground state since the singlet state of the deuteron cluster has a higher energy than the triplet state. From arguments similar to those applied to Li⁷ one sees immediately that all Li⁶ states considered here have even parity. These qualitative conclusions agree with the experimental level scheme in Fig. 3, see also [30].

Quantitative considerations follow later.

As third example we consider the low energy levels of $O^{16}$. We discuss this example, particularly the low excited negative and positive parity states, in somewhat greater detail because it indicates especially clearly how we must generalize the too simple oscillator cluster functions so as to improve our variational calculations. In order to begin our description of the low energy $O^{16}$ states in the oscillator cluster model we assume an $\alpha$-cluster $C^{12}$-cluster structure. This is equivalent to a representation in terms of four $\alpha$-clusters, if $C^{12}$ is represented by three $\alpha$-clusters without internal excitation, R. SHELINE and K. WILDERMUTH [33].

In the groundstate of $O^{16}$ the relative motion of the unexcited $\alpha$-cluster about the unexcited $C^{12}$-cluster has four oscillator quanta of energy and zero angular momentum. This follows, as we emphasize again, because the resulting (antisymmetrized) wave function must be mathematically equivalent to the usual oscillator shell model wave function of the $O^{16}$ ground state, where the $1s$ and $1p$ shells are completely filled.

We now consider the lowest negative parity excited levels. In these states when the $\alpha$- and $C^{12}$-clusters are not internally excited, the relative motion of the two clusters corresponds to five oscillator quanta of energy and hence to orbital angular momentum 1, 3 or 5. We obtain further negative parity levels when the $C^{12}$-cluster is internally excited to its first $(2^+)$ or second $(4^+)$ excited state⋆; the spins of the internally excited $C^{12}$-clusters will be coupled to the above mentioned orbital angular momentum values. All these negative parity oscillator cluster functions have just one oscillator quantum of energy above the groundstate oscillator cluster functions.

After antisymmetrization most of the negative parity wave functions for $O^{16}$ described above (the coupling possibilities range from $I = 1^-$ to $I = 9^-$) either vanish or become identical to one another. To see this we employ our usual method of considering the levels simultaneously in two different representations: the single-particle oscillator shell model and the oscillator $\alpha$-cluster $C^{12}$-cluster representation. In the shell model

---

⋆ In these two excited $C^{12}$ states no $\alpha$-cluster is internally excited and, in terms of the oscillator shell model, all nucleons remain in the lowest single particle states compatible with the Pauli principle (i. e. no excitation from the $1p$ to the $2s$ or $1d$ shell). This is quite analogous to the corresponding $2^+$ and $4^+$ states of Be⁸ discussed earlier [see text following Eq. (III, 12)].

representation all these negative parity levels of $O^{16}$, having one oscillator quantum of excitation energy and in which none of the four $\alpha$-clusters are internally excited, correspond to one particle excitations to the $1d, 2s$ shell (see App. B). One can have only such one particle excitations in which no spin flip occurs, because otherwise an $\alpha$-cluster would be broken up. But this means that the total spin of these states is formed by the orbital angular momenta of the nucleons. By lifting up one nucleon from the $1p$ oscillator shell to the $2s, 1d$ oscillator shell the orbital angular momenta of the nucleons can form only the spins 3, 2, 1. Further one sees that for spin 3 and also for spin 2 just one coupling possibility exists. Therefore $O^{16}$ has one $3^-$ and one $2^-$ state, which are one-particle excitation states to the next higher oscillator shell and in which simultaneously no $\alpha$-cluster is broken up. For the spin 1 there are two such coupling possibilities. But one of these couplings must describe a pure center of mass excitation of the $O^{16}$ nucleus in the space fixed oscillator potential. The orbital angular momentum of this excited center of mass motion is $l = 1$. This is easily seen by briefly referring back to the oscillator cluster picture. There we can obtain from the ground state a negative parity state with one oscillator quantum of excitation by raising the (specious) center of mass function from a $1s$ to a $1p$ oscillation. Because the center of mass coordinate is symmetric in the particle coordinates, this specious state cannot vanish upon antisymmetrization. Hence there also exists only one spin 1 negative parity $O^{16}$ state in which a single particle is excited to the next oscillator shell and no $\alpha$-cluster is broken up. One sees from this that from all spin-coupling possibilities which one expects in the $C^{12}$-$\alpha$ oscillator cluster representation $(I = 1^- \ldots I = 9^-)$ only three possibilities remain at the most. This reduction is due to the Pauli principle. This is an especially nice example of how it is sometimes possible to take into account the effect of the Pauli principle without explicit calculations by considering the states of a nucleus simultaneously in different representations and retaining only those states which are compatible with all of these representations. These kinds of considerations are also very helpful in constructing the appropriate nuclear wave functions explicitly. For instance, in the case of $O^{16}$ as soon as one has constructed one $1^-$ wave function for a negative parity state with one oscillator quantum of excitation and no broken-up $\alpha$-cluster one has found already all possible wave functions of this kind due to our previous consideration. The same is true for the corresponding $2^-$ and $3^-$ states*.

As one expects from our considerations, the lowest measured negative parity levels of $O^{16}$ have the spins $3^-$, $1^-$ and $2^-$. As one might expect, the $2^-$ level lies above the $1^-$ and $3^-$ levels, because there the $C^{12}$-cluster has to be in its first excited state. The penetrating orbit argument no longer suffices to give the ordering of the $1^-$ and $3^-$ levels, since there is large cluster overlapping in both these states and other more complicated effects connected with the Pauli principle come into play. If one calculates

---

* For the explicit construction of these wave functions see [25].

the spacing between the $I = 1^-$ and the $I = 3^-$ level using oscillator cluster trial functions with the Serber Ansatz (III,14), then one finds that these two levels coincide*. The spin-orbit part of the nuclear forces introduces a small probability that an $\alpha$-cluster will break up, and thereby depresses the state with the larger angular momentum; therefore the first $3^-$ state in $O^{16}$ has a little smaller excitation energy than the first $1^-$ state.

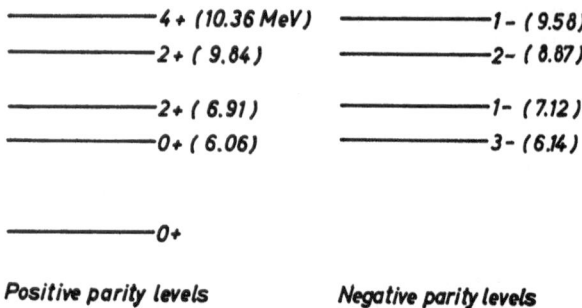

Positive parity levels                 Negative parity levels

Fig. 4. Level scheme for $O^{16}$

The first calculations of the lowest negative parity states of $O^{16}$ with $T = 0$ were made in the frame of the oscillator shell model (one oscillator quantum excitation), J. ELLIOT and B. FLOWERS [34]. These calculations give the correct energy sequence of the three lowest states practically independently of the forces postulated, whereas the other $T = 0$ negative parity states do not fit in with the experimentally measured states, [30]. The reason for this follows immediately from our previous considerations. The three first negative parity states coincide with the three states which we have considered above (if one neglects spin-orbit coupling)**. Due to our considerations, in the other negative parity states of Elliot and Flowers an $\alpha$-cluster must always be broken up. Therefore their excitation energies should be at least 12 MeV or higher instead of, for instance, 9.58 MeV for the fourth negative parity state of $O^{16}$. It is energetically much easier to excite a higher relative motion of the outside $\alpha$-cluster against the $C^{12}$ ground state cluster than to break up an $\alpha$-cluster. Therefore one has to assume that the fourth negative parity state of $O^{16}$ is again approximately an $\alpha$-particle state, but with a relative cluster oscillation of higher order. In the frame of the oscillator model, this negative parity state must have an excitation of seven oscillator quanta. For the lowest $\alpha$-particle state of seven oscillator quanta excitation one expects a $4p$ oscillation of the $\alpha$-cluster relative to the unexcited $C^{12}$-cluster (see App. B). The spin of this state must be $I = 1^-$, which agrees with the spin of the fourth negative parity level (9.58 MeV) of $O^{16}$. Similar considerations are also valid for the succeeding negative parity states.

---

* Unpublished calculation of TH. KANELLOPOULOS.

** It is interesting to mention that these states and the $O^{16}$ ground state also coincide with the corresponding zero approximation particle-hole states, see for instance G. BROWN [35]. This is again due to the indistinguishability of the nucleons.

We shall now discuss the low-excitation levels of $O^{16}$ having positive parity; these are also $\alpha$-particle levels. In the frame of the oscillator cluster model the lowest excited positive parity level ($I = 0^+$) is described by a *4s* oscillation (6 oscillator quanta of excitation) of the $\alpha$-cluster relative to the unexcited $C^{12}$-cluster. This $0^+$ state at 6.06 MeV excitation energy is the lowest level in a rotational band whose next levels are $2^+$ at 6.9 MeV excitation energy and $4^+$ at 10.4 MeV excitation energy*. Above the rotational $2^+$ state lies a second $2^+$ state at 9.84 MeV; the relative motion of the $\alpha$-cluster is again a *4s* oscillation, but the $C^{12}$-cluster is internally excited to its first excited ($\alpha$-particle) state ($I = 2^+$). This description of the 9.84 MeV level is supported by the fact that its energy distance from the first excited $I = 0^+$ level is 3.8 MeV, which is approximately the excitation energy of the first excited $C^{12}$ level (4.4 MeV).

If one calculates the energy expectation values of all the $O^{16}$ levels considered so far in the frame of the oscillator shell or oscillator cluster model then one obtains much too high excitation energies. With the forces (III,14) for instance one obtains for the negative parity levels excitation energies of 15 MeV and more and for the positive parity levels 20 MeV and more**. This is in complete disagreement with the experimental values. There the lowest energy level has an excitation energy of 6.06 MeV. Further the lowest excited positive parity level has a lower excitation energy than the lowest negative parity level. In the frame of the oscillator cluster model one expects just the opposite because the lowest excited positive parity level has 6 oscillator quanta of excitation and the lowest negative parity level has 5 oscillator quanta of excitation.

Therefore one has to ask: what causes the low excited states of $O^{16}$ to have much smaller excitation energies than one would expect from the above calculations? The main reason, as we shall discuss in the next section, is that the different relative and internal motions of the clusters have different oscillation frequencies, contrary to the oscillator cluster model where all these frequencies are equal to each other. Here we shall only mention that this effect is of quite general nature and therefore influences more or less all nuclear states. The effect goes always in the direction that the energies of the nuclear states are lowered, often 10 MeV and more as in the above discussed excited states of $O^{16}$.

Finally, before leaving the oscillator cluster model, we wish to make some additional remarks on the problem of choosing the most suitable oscillator cluster function expansion system. We have already emphasized that each of these expansion systems is complete, so that our choice is a matter of convenience in simplifying the calculations and improving the physical insight. The important point we make here is that our choice not only depends on what nuclear level we are considering but also on the kind of nuclear reaction process we wish to describe. For instance in $O^{16}$ transition probabilities from excited states with no broken-up

---

* See also Chapter IV section E on rotational states.

** In these calculations the width-parameter of the oscillator potential has to be fixed to take into account in rough approximation the saturation character of the nuclear forces as we have discussed earlier.

$\alpha$-cluster, the $\alpha$-cluster description discussed previously is the most appropriate for the $O^{16}$ ground state. However in the giant resonance state of $O^{16}$ the description is, to a first approximation, that the protons as a whole and the neutrons as a whole are in a relative dipole oscillation state. The most convenient description of this dipole motion is a cluster of eight protons and another cluster of eight neutrons, with an oscillatory relative motion $\chi_k(R_{8n} - R_{8p})$ between the two clusters, [10]. The oscillator cluster wave function then is given by

$$\psi = \mathscr{A} \{\phi_i(8n)\, \phi_j(8p)\, \chi_k(R_{8n} - R_{8p})\}. \qquad \text{(III,32)}$$

For $k = 0$, or zero order oscillation between the neutron and proton clusters, we get the ground state of $O^{16}$, that is, in the oscillator cluster model, the antisymmetrized eight proton–eight neutron oscillator cluster function with no oscillator quanta in the relative motion of the clusters is identical to the antisymmetrized shell model (and $\alpha$-cluster) ground state-description. This is the most convenient description of the $O^{16}$ ground state in dealing with transitions to the $O^{16}$ giant dipole state.

Another example is $Li^6$. We can describe its ground state and low excited states either by an $\alpha$-cluster and $d$-cluster, or by a *triton*-cluster and $He^3$-cluster. In the oscillator model these descriptions are mathematically identical under antisymmetrization. We shall discuss this example in more detail later.

We emphasize that such simple equivalences hold strictly only in the oscillator cluster model description; indeed this is one of its advantages. In the generalization of the cluster functions, to which we now turn, we begin to strengthen certain cluster correlations at the expense of others, and these simple equivalences become only approximate.

## IV. Generalization of the Cluster Wave Functions

### A. Nuclear bound states

So far in considering cluster wave functions we have set the oscillator frequencies to be always the same for the internal motion of the nucleons within the clusters and for the relative motion of the clusters. This is the same as saying that the nucleons move in a common oscillator potential. This assumption leads to excessively large level spacing, for example in the case of $O^{16}$, in contradiction to experience. This means that, so long as we restrict ourselves to the superposition of a small number of oscillator cluster functions for the description of nuclear states, the oscillator cluster model can very often give us good insight into the qualitative structure of nuclear spectra, but usually can not give good quantitative results.

To improve the original oscillator cluster functions we introduce additional variational parameters into these wave functions. This generalization of the cluster wave functions must be such that under the Ritz variational minimization the energies of the levels decrease as much as possible. As mentioned already in our discussions of $O^{16}$ the most important variational parameters, which we allow to vary independently of each other in the different states, are the internal frequencies of the clusters and the frequencies of their relative motions. Inspection of the cluster functions shows that the internal frequency parameter of a cluster determines the strength of that cluster correlation. The larger the internal frequency the greater the probability that the nucleons in the cluster must be found closer together. More roughly we can say the internal frequency parameter determines the mean radius of the cluster (insofar as we can speak of a cluster having a radius); the greater the internal frequency the smaller the mean radius of the cluster. Similarly the relative motion frequency parameter determines roughly the mean separation of the cluster centroids. Hence we shall alternatively refer to these frequencies respectively as the internal and relative motion width parameters.

The important fact we are stating is that by strengthening certain cluster correlations and adjusting their separation, which of course is not directly possible in the simple oscillator cluster model or shell model, we can greatly reduce the energy expectation value $\langle H \rangle$ and obtain good quantitative agreement with experiment. This is the justification of the whole physical picture of cluster correlations. In addition we shall see that the generalized cluster functions provide a logical connection between the theories of nuclear structure and nuclear reactions.

The introduction of such generalized cluster functions means a very strong deviation from the usual shell model picture because now it is no longer possible to define a common average potential in which all nucleons of a nucleus move*. To understand the physical meaning of these new degrees of freedom in the variational functions let us again consider the special example $O^{16}$. If one compares the frequency (or width) parameter, $a = M\omega/\hbar$, of a free $\alpha$-particle with the frequency parameter of the $O^{16}$ ground state in the oscillator cluster model, then one finds that the former is larger**; this means that a free $\alpha$-particle has a larger frequency parameter (i.e. smaller mean radius) than an $\alpha$-cluster in $O^{16}$, [24]. In the excited states at least some of the $\alpha$-clusters are further outside of the nucleus and therefore they behave more like free $\alpha$-particles. This means their internal frequency parameter should increase. Further, with increasing excitation energy the frequencies of the

---

* It is certainly also possible to expand such generalized cluster functions in terms of oscillator cluster functions, because they too form complete sets of wave functions. But this would mean that, as in the shell model picture, one has to include in such expansions many very highly excited oscillator cluster functions. Therefore such expansions become very complicated and one loses any physical insight.

** One calculates $\langle r_i^2 \rangle$ of $O^{16}$ and $He^4$ and fixes the parameters $a$ of $O^{16}$ and $He^4$ in such a way that these $\langle r_i^2 \rangle$ values coincide with the corresponding experimental $\langle r_i^2 \rangle$ values.

relative motions of the clusters in general decrease. This is due to the effect that in the limit case where one $\alpha$-cluster becomes a free particle this frequency must go to zero, an anharmonicity effect. Let us consider a quantitative example.

We consider the ground state and the two first excited states of $He^5$ and $Li^5$ (see Fig. 5). The ground state of $He^5$ or $Li^5$ is considered as an $\alpha$-cluster and a nucleon; therefore for the trial wave function of $He^5$ we make the following "ansatz", L. D. PEARLSTEIN et al. [36]:

$$\psi(He^5) = \mathscr{A}\{\phi_0(a)\,\phi_0(n)\,\chi(a-n)\}$$

$$= \mathscr{A}\left\{e^{-\frac{a}{2}\sum_{i=1}^{4}\bar{r}_i^2}\cdot R\,e^{-\frac{2}{5}\gamma R^2}\,Y_{l,m}(\theta_R,\,\varphi_R)\,\xi(s_1\ldots t_5)\right\} \qquad (IV,1)$$

where

$$\bar{r}_i = r_i - R_1;\ i = 1, 2, 3, 4$$

$$R_1 = \frac{1}{4}\sum_{i=1}^{4} r_i$$

$$R = R_1 - r_5\ .$$

The spin and isobaric spin function, $\xi(s_1\ldots t_5)$, after antisymmetrization describes an $S = \frac{1}{2}$, $T = \frac{1}{2}$ state.

For $a = \gamma$ the wave function (IV,1) goes over into the usual oscillator shell model wave function of $He^5$ if one splits off there the center of mass motion of $He^5$. In the oscillator shell model one sees that the wave function (IV,1) will vanish unless there is at least one oscillator quantum of excitation. Since we consider only unexcited clusters here this means the lowest relative motion function must be a $1p$ function. When we generalize the cluster function (IV,1) by using the width parameters as free variational parameters, the angular dependence of the relative motion function remains unaltered. Hence the lowest states will have

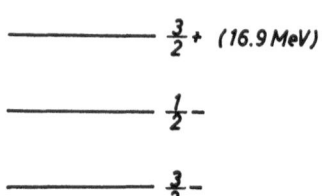

$$\frac{3}{2}+\ (16.9\,MeV)$$

$$\frac{1}{2}-$$

$$\frac{3}{2}-$$

Fig. 5. Level scheme for $He^5$

orbital angular momentum $l = 1$. Because the unexcited $\alpha$-cluster and the neutron have positive internal parity, these low $l = 1$ states will have negative parity.

Since we have not yet used nucleon-nucleon forces that give saturation we must employ our usual trick of putting in the saturation effect phenomenologically, just as we did previously for $Be^8$ in the oscillator cluster model. From the arguments following eq. (III,14) we saw that while the hard-core forces were not expected to contribute appreciably to the energy of light nuclei, they did serve to prevent the collapse of the nucleus. This means the hard-core forces (together with the Pauli principle) put an upper limit on the strength of any cluster correlation, or more roughly, due to the saturation character we cannot decrease the "radius" of a cluster indefinitely. Hence if we neglect hard core forces we must estimate the $a$-cluster radius in a phenomenological way. Here we assume the radius of the $a$-cluster is not affected by the outside nucleon. Thus we fix $a$ such that the $a$-cluster has the same radius (internal frequency) as a free $\alpha$-particle, which means at the moment we can discuss only the second kind of generalization (anharmonicity effect). $\gamma$ is a completely free variational parameter. Later, when we employ hard core forces, the internal width parameters also become free variational parameters and our results are correspondingly improved.

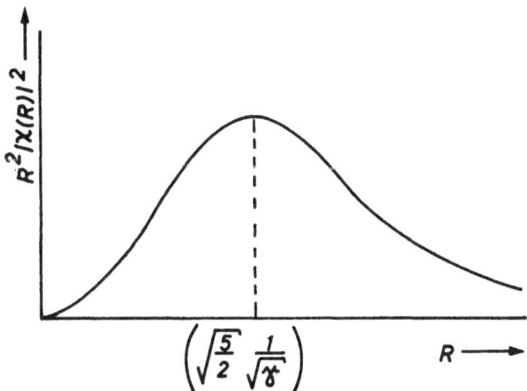

Fig. 6. Trial relative motion function for He⁵ ground state

The wave function of the relative motion is shown in Fig. 6 for the $\frac{3^-}{2}$ ground state, where the peak at $R = \sqrt{\frac{5}{2}\left(\frac{1}{\gamma}\right)}$ is the most probable distance of the neutron from the center of mass of the $a$-cluster.

The Hamiltonian for He⁵ is given by

$$H = \sum_{i=1}^{5} \frac{p_i^2}{2M} + \sum_{i=1}^{5} \sum_{j>i}^{5} V_{ij} \qquad \text{(IV,2)}$$

where for $V_{ij}$ we must add a spin-orbit term to the simple Serber ansatz (III,14). To see that the spin-orbit term is necessary here for a quantitative description, we observe that the two lowest states of He⁵ have the same orbital angular momentum, which in accordance with the Pauli principle will be $l = 1$. The outside neutron can couple with $l$ to form the two states $\frac{3^-}{2}$ and $\frac{1^-}{2}$. The splitting of these levels is therefore due to the spin-orbit coupling and thus depends on the total angular momentum. Unfortunately it is not yet possible to determine a phenomenological spin-orbit force from the two-nucleon scattering data, so we must resort to the less desirable method of analysing the nuclear spectroscopic data. We choose for the form of the spin-orbit force the simplest translation invariant "ansatz" without exchange terms, namely:

$$V_{ij}^{LS} = -\frac{V_0^{LS}}{2\hbar} \exp\left\{-\beta\left(r_i - r_j\right)^2\right\} \cdot \left(r_i - r_j\right) \times \left(p_i - p_j\right) \cdot \left(\sigma_i - \sigma_j\right) \qquad \text{(IV,3)}$$

where $V_0^{LS}$ and $\beta$ are parameters to be phenomenologically determined. To reduce the number of free parameters which have to be fixed from experimental data the range parameter $\beta$ was chosen the same as in the central two body potential (III,14)*. The only parameter which remains to be fixed is $V_0^{LS}$.

$V_0^{LS}$ can be phenomenologically determined from the spin-orbit splitting of the $\frac{1}{2}^-$ and $\frac{3}{2}^-$ levels of the Li$^7$ nucleus. The structure of these levels is in good approximation that of an $\alpha$-cluster plus a *triton*-cluster. The spin of the $\alpha$-cluster is zero and the spin of the $t$-cluster is $\frac{1}{2}$, the orbital angular momentum is $l = 1$ for both Li$^7$ states (which means their parity is negative). In the $\frac{3}{2}^-$ level $\left(\frac{1}{2}^- \text{ level}\right)$ the $t$-cluster spin and orbital angular momentum are parallel (anti-parallel) to each other. Therefore in this approximation the energy splitting of the two levels comes only from the spin-orbit coupling (IV,3). For $V_0^{LS}$ one then obtains 3.8 MeV. Practically the same value is obtained by using oscillator cluster functions or generalized functions of the kind just described. In the following calculations we will use (IV,3) with the parameter values discussed here for the spin-orbit term of the phenomenological two nucleon potential, [29].

The expectation value of the energy of He$^5$ is now found as function of $1/\sqrt{\gamma}$. The result of the calculation is indicated schematically in Fig. 7.

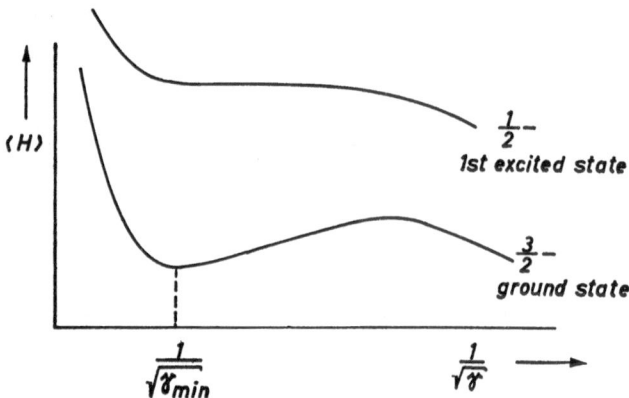

Fig. 7. Schematic of the energy expectation value vs. $\sqrt{1/\gamma}$ for He$^5$

Let us try to understand the result of Fig. 7. As $\gamma$ increases, the average distance between the neutron and the $\alpha$-cluster decreases. Therefore due to the uncertainty principle the relative kinetic energy between the nucleon and the $\alpha$-cluster increases. For $1/\sqrt{\gamma}$ approximately

---

* It turns out, for instance by looking at the spin-orbit splitting of the Li$^6$ levels, that this assumption is not very good. But for the arguments presented here it does not matter very much. Work is in progress to find a more satisfactory expression for the two-nucleon spin-orbit force.

equal to the $\alpha$-cluster "radius" the neutron comes, with a large probability, within the range of the $\alpha$-cluster. Therefore, due to the short
range character of the nuclear forces, the negative potential interaction
energy increases by a large amount, and turns the trend of the energy
curve downward. For large $\gamma$ values the neutron is completely inside the
$\alpha$-cluster and the interaction potential energy remains more or less
constant, while the kinetic energy continues—due to the uncertainty
principle—to increase.

When $\gamma$ is equal to $\gamma_{min}$, and the nucleus is in a virtual state, the
energy $\langle H \rangle$ has a relative minimum. We observe the ground state is a
$\frac{3^-}{2}$ state in agreement with experiment. For the decay of He$^5$ into an
$\alpha$-particle plus a free neutron, the neutron must tunnel through the
resultant barrier as indicated in Fig. 7. The barrier will therefore be
related to the lifetime of the width of the state*.

The first excited state of He$^5$, which is separated from the ground
state by the spin-orbit force, has a short lifetime and a large width,
since the relative minimum in $\langle H \rangle$ is very shallow due to the repulsive
character of the spin-orbit force in this state.

The 16.9 MeV $\frac{3^+}{2}$ excited state of He$^5$ cannot be explained by an
$\alpha$- and *neutron* cluster structure since it is a resonance level with a large
life time, and it can be shown that for a $\frac{3^+}{2}$ state of this cluster structure
$\langle H \rangle$ as a function of $1/\sqrt{\gamma}$ has no relative minimum. For this level of
He$^5$ we assume instead a *triton* plus *deuteron* cluster structure, W. Kunz
[37, 36]. It is seen immediately that this cluster structure has a much
higher excitation energy than the $\alpha$- plus $n$-cluster structure, since the
$\alpha$-cluster is broken up and the $d$-cluster has a small binding energy. We
can repeat the calculation for a state with this cluster structure. The
wave function of the 16.9 MeV excited state of He$^5$ we approximate due
to our assumption as follows, [36]:

$$\psi(He^5, 16.9 \text{ MeV}) = \mathscr{A} \, \phi(\underset{\sim}{t}) \, \phi(\underset{\sim}{d}) \, \chi(t - d) \qquad (IV,4)$$

or in more detail

$$\psi = \mathscr{A} \exp\{-(a/2) \sum_{i=1}^{3} \overline{r}_i^2\} \exp\{-(b/2) \sum_{i=4}^{5} \overline{r}_i^2\} R^n \exp\{-(3/5)\,\beta R^2\} \times \qquad (IV,5)$$
$$\times \, Y_{em}(\theta_R \, \varphi_R) \, \xi(s_i t_i)$$

with

$$\boldsymbol{R} = \boldsymbol{R_t} - \boldsymbol{R_d} .$$

$\xi(s_i t_i)$ describes again the spin and isobaric spin state of the excited
He$^5$ state $S = \frac{3}{2}$, $T = \frac{1}{2}$. $\beta$ and $b$ are taken as the variational parameters
and the radius of the cluster is fixed**. We choose $n = 2$ and $l = 0$ to get
the smallest excitation energy. $l = 0$ means we have a maximum mutual

---

* However see the further remarks on this point below.
** Again due to the non-saturation character of the nuclear forces used in these
calculations.

penetration of the $t$- and $d$-cluster. For $a = b = \beta$ the wave function (IV,5) becomes equal to the corresponding oscillator cluster function. The internal parity of the $t$- and $d$-cluster is positive; therefore the total parity of the wave function (IV,5) is positive due to the even angular momentum of the relative motion of the two clusters. The spin of the $t$-cluster is $\frac{1}{2}$ and that of the $d$-cluster is 1. The lowest energy value, due to the spin-exchange character of the nuclear forces, is for the case of parallel spins, and thus the lowest state above the ground state is a $\frac{3^+}{2}$ state. This is in agreement with the experiments.

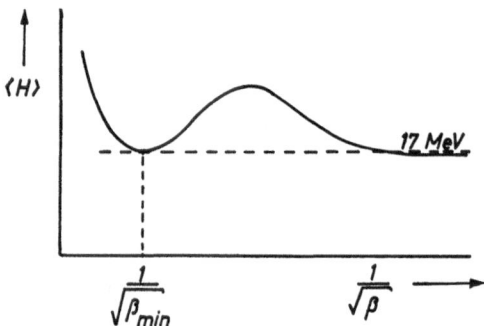

Fig. 8. Schematic of the energy expectation value vs. $\sqrt{1/\beta}$; the energy $\langle H \rangle$ is minimized with respect to $b$ for each value of $\beta$

The results of the calculation of $\langle H \rangle$, sketched in Fig. 8, show a very deep relative minimum, indicating that a resonance level with a long half-life (and therefore a small energy width) should exist. For $\beta = \beta_{min}$ we get the excitation energy of this level as $\approx 17$ MeV. This proves theoretically that the narrow 16.9 MeV $\frac{3^+}{2}$ state of He⁵ has a $d$- plus $t$-cluster structure. The decay probability of this level into an $\alpha$-particle and a neutron is very small, as can be shown by a rough estimation, because the decay can be induced only by a spin-orbit or tensor force. Note for the oscillator representation where $a = b = \beta$, one obtains an excitation energy of about 23 MeV. Thus by introducing $b$ and $\beta$ as free variational parameters, which introduces a strong deviation from the oscillator assumption or the usual shell model descriptions with interaction, we get a better wave function for the 16.9 MeV $\frac{3^+}{2}$ level of He⁵ and a depression in its excitation energy of about 6 MeV. The description of Li⁵ levels (mirror levels to He⁵) goes completely analogously.

As one sees without further explanation, the deviation from the oscillator assumption will have very similar consequences for the excitation energies of all nuclear states. For instance in the case of O¹⁶ the deviation from the oscillator assumption and therefore the just described energy depression effect will be larger for the excited positive parity $\alpha$-particle states than for the negative parity $\alpha$-particle states. Together with the cluster-radius change effect, which we have not yet completely discussed quantitatively, this makes it understandable why the first excited state of $O^{16}$ is a positive parity $\alpha$-particle state and not a negative parity $\alpha$-particle state.

The phenomenon of very narrow levels suddenly following above very broad levels is found in many nuclei in addition to $He^5$ and $Li^5$, see Fig. 9. The explanation is always the same as in the present case, namely that the transition marks a change in cluster structure. For instance, in $Be^8$ one has after the very broad $4^+$ level many very narrow levels in the energy region between 16 and 20 MeV. In these higher states one of the $\alpha$-clusters is broken up into a $t$-cluster plus a proton or a $He^3$-cluster plus a neutron. An adequate quantitative description requires a superposition of both kinds of states, namely an $\alpha$-$t$-$p$ cluster state plus an $\alpha$-$He^3$-$n$ cluster state★ (that is, we must use at least these two terms in the cluster function expansion of our trial function to obtain a satisfactory approximate solution from the Ritz variational principle). Since two clusters may always be combined to form a larger

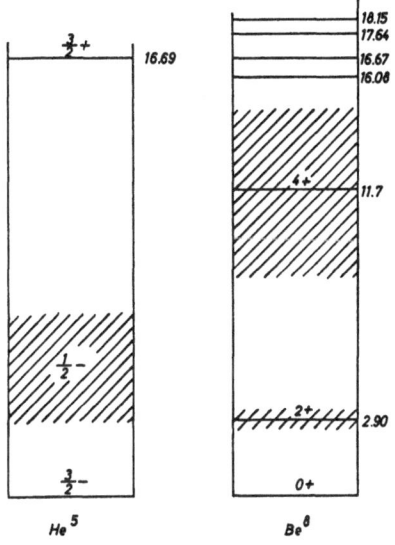

Fig. 9. Examples of spectra indicating change of cluster structure

cluster, we may also denote these $Be^8$ excited states as a superposition of a $Li^5$-$t$ cluster state plus a $He^5$-$He^3$ cluster state, or equivalently under antisymmetrization, a $Li^7$-$p$ cluster state plus a $Be^7$-$n$ cluster state. The new clusters have a much stronger mutual interaction than two $\alpha$-clusters. Hence in order to decay into the corresponding free particles, these clusters must overcome a considerable potential barrier. Therefore, even though they have a higher excitation energy than the $Be^8$ $\alpha$-cluster states, the new cluster states have a narrower level width★★.

This explanation for the fact that in nuclei very often with increasing excitation energy after broad levels suddenly again narrow levels appear looks very natural and very "anschaulich". However the explanation is

---

★ Due to the charge independence of the nuclear forces these two states must have equal amplitudes if we neglect the Coulomb force.

★★ It can even happen that the effective interaction energy between the clusters, i. e. the total energy minus the internal energy of the clusters, is large enough to form a bound cluster state (by bound state we mean the curve of interaction energy vs. relative motion width-parameter possesses an absolute minimum instead of a relative minimum). Such bound cluster states cannot decay into the corresponding free particles. An example occurs for the 16.08 MeV $Be^8$ level. The energy width of this state is essentially defined by the decay probability into the two $\alpha$-particle channel.

not completely correct, because the potential barriers of Fig. 5 and 6 follow from the special form of our trial wave functions. The potential barriers do not appear in such a way in nuclear reactions produced by incident beams. After we have introduced the nuclear reaction theory we shall present some additional remarks to make the explanation correct.

As next example of the use of generalized cluster functions, we consider the ground state of $Li^6$. As will be shown in Appendix C, in the oscillator model the ground state of $Li^6$ can be described by the following equivalent functions, Y. TANG et al. [38, 32]:

$$\psi(Li^6, \text{ground state}) = \mathscr{A} \{\phi_0(\underset{\sim}{a})\; \phi_0(\underset{\sim}{d})\; \chi(a - d)\}$$
$$= N \mathscr{A} \{\phi_0(\underset{\sim}{t})\; \phi_0(He^3)\; \chi(t - He^3)\}\,. \tag{IV,6}$$

$\chi(a - d)$ and $\chi(t - He^3)$ are oscillator wave functions of two oscillator excitation quanta and with zero orbital angular momentum. $\phi_0(a)$, $\phi_0(d)$, $\phi_0(t)$ and $\phi_0(He^3)$ are the internal functions for the ground state representations of an $a$-, $d$-, $t$-, and $He^3$-cluster respectively in the oscillator frame. The spins of the $t$- and $He^3$-cluster are parallel to each other. The normalization constant $N$ is necessary to make the two oscillator cluster functions of (IV,6) equal to each other. To leave the oscillator model one chooses as before the width parameters of the constituent wave functions in the equations above independently of one another. As a consequence the equivalence of the two cluster wave functions is lost. With the help of the Ritz variational method one can easily show that a generalized $a - d$ cluster wave function yields a better description of the ground state of $Li^6$ than a generalized $t$-$He^3$ cluster wave function. This is to be expected, because if one expands the generalized $t$-$He^3$ cluster wave function in the $a$-$d$ cluster function system, one finds that it contains a part which describes a broken-up $\alpha$-cluster, the amplitude of which increases with increasing deviation from the oscillator assumption. Thus it is energetically very unfavorable for the $t$-$He^3$ cluster wave function to deviate appreciably from its form in the oscillator model. The variational calculation bears out our argument: by using the relative motion width parameter of the respective cluster functions as variational parameter and minimizing the respective energy expectation values one obtains a much stronger deviation from the oscillator assumption and a much larger binding energy in the case of the $a$-$d$ cluster wave function than in the case of the $t$-$He^3$ cluster wave function. Physically, this means that the asymptotic behavior of this state can be better expressed in terms of an $\alpha$-particle plus a deuteron than a triton plus a $He^3$ nucleus. (By asymptotic behavior, one means the behavior of the wave function for large distances between the clusters. Here the effect of the Pauli principle vanishes asymptotically). A good indication of the validity of our arguments is found from the experimental results that break-up of the $Li^6$ ground state into a free $\alpha$-particle and deuteron requires only 1.5 MeV, whereas break-up into a free t plus $He^3$ requires some 15 MeV. This bears out our previous remark on the importance of cluster correlations near the nuclear surface.

Again the above example shows the general property that the difference between two different cluster wave functions increases as one goes away from the oscillator model; the properties of the cluster approach those of the corresponding free particles as the deviation from the oscillator assumption becomes larger. Nevertheless, one must still remember that even in the latter case the indistinguishability of the nucleons, quantitatively expressed by their antisymmetrization, still has the effect of reducing the differences between different generalized cluster wave functions.

In the above example we have seen that, due to the high internal binding energy of an $\alpha$-cluster, $Li^6$ behaves asymptotically more like an $\alpha$-particle plus a deuteron than like a triton plus a $He^3$. This result has some interesting general implications. If we consider medium or heavy nuclei in their ground states, then the wave functions of these states can be written as a superposition of wave functions of any antisymmetrized complete cluster wave function system. For instance, these states can be described as a relative motion of an $\alpha$-cluster, $t$-cluster, $He^3$-cluster, and so on (including internal excitation of the cluster), against the corresponding residual clusters. The amplitude of the relative motion will be especially large if the internal binding energy of the clusters is especially big. Therefore near the nuclear surface usually one will "see" with a relatively large probability $\alpha$-clusters, with a smaller probability for $t$-clusters and a yet much smaller probability (especially in heavy nuclei) for $He^3$-clusters, see for instance D. WILKINSON [39]. The latter follows from the fact that it is energetically very unfavorable to remove protons from a heavy nucleus because this increases the positive asymmetry energy. In the case of $\alpha$-clusters the high internal binding energy of these clusters overcompensates this tendency.

## B. Characteristics of generalized cluster wave functions

With just the oscillator cluster model assumption it is possible by rather simple arguments to qualitatively predict the sequence of energy levels for many light nuclei, although the quantitative results are usually poor. By introducing the internal and relative motion widths as independent variational parameters we were able to obtain a large decrease in the minimum of the Ritz variational principle. For example, for the 16.9 MeV level of $He^5$ the decrease in the energy expectation value from the oscillator cluster model or oscillator shell model value was 6 MeV; a rough estimate for the excited levels of $O^{16}$ yields a corresponding decrease of as much as 10 MeV. Moreover in all the nuclear levels treated to date the calculated energies were in quite reasonable accord with experiment, so that our theoretical picture of cluster states is now capable of quantitative results. But this generalization means a considerable departure from the simple oscillator view. We must now inquire which characteristics remain and which must be modified when we generalize an oscillator cluster function [24, 32, 33], K. WILDER-MUTH [40]. We also wish to point out how flexible and general generalized cluster functions really are.

The most important properties of oscillator cluster functions unaffected by generalization are the total orbital angular momentum, the spin, the isospin and parity. This is apparent because the width parameters affect only the radial parts of the cluster functions and not the angular or spin dependence. Therefore if an oscillator cluster function is an eigenfunction of total angular momentum and parity, the corresponding generalized cluster function will also be an eigenfunction to the same eigenvalue. The separation of the total center of mass motion function is likewise unaffected by generalization, since by the introduction of cluster coordinates it is split off from the beginning. Because the antisymmetrization operator commutes with the total orbital angular momentum, total spin and isospin, and parity operators, and because the center of mass coordinates of the nucleons are symmetric with respect to all nucleon coordinates, these general properties are not altered by antisymmetrization of the generalized cluster wave function.

On the other hand certain properties of oscillator cluster functions become modified as we depart from the oscillator assumption. The most important of these is the distinction between different cluster descriptions of the same nuclear level. We have seen for instance in the oscillator model the ground state of $O^{16}$ can be described either by an eight proton-eight neutron cluster function, or by a $C^{12}$-$\alpha$ cluster function and so on. These wave functions are identical upon antisymmetrization. Similarly, upon antisymmetrization the $\alpha$-$d$ oscillator cluster function and $t$-$He^3$ oscillator cluster function are identical wave functions describing the low energy levels of $Li^6$ (see App. C for explicit proof). The equivalence of different cluster functions under antisymmetrization is true for many states of nuclei, so long as we stick to the oscillator assumption. However, this equivalence begins to be lost as we permit the internal and relative motion widths to depart from their common oscillator value. This means that we strengthen certain cluster correlations at the expense of others, so that the distinction between the various cluster descriptions becomes greater the further we go from the oscillator model, and there is usually a single best cluster description for any nuclear level. Nevertheless so long as the deviation from the oscillator assumption is not too great (which is the case for many nuclei) these equivalences still exist approximately in the sense that the mutual overlap of different cluster descriptions of the same nuclear level is often large. We emphasize that the mutual overlap would not be nearly close to unity if the different generalized cluster functions were not antisymmetrized.

A second characteristic requiring modification when we generalize the oscillator cluster functions can be pointed out by the following example. In the oscillator model we have seen that due to the Pauli principle no $5^-$ $\alpha$-particle state exists among the low excited negative parity levels of $O^{16}$. In other words, an oscillator cluster function describing a $C^{12}$- and an $\alpha$-cluster in their ground state and with a relative motion corresponding to five oscillator quanta and orbital angular momentum $5^-$ vanishes under antisymmetrization. This is certainly no more true if we introduce the width parameter of this relative motion as additional variational parameter. However, the energy expectation value for such a state will be very high. To see this, one expands the generalized $5^-$ cluster function in $C^{12}$-$\alpha$ oscillator cluster functions. Due to the Pauli principle the lowest oscillator cluster function in this expansion will describe an $\alpha$-cluster with relative motion about the $C^{12}$-cluster of 7 oscillator quanta excitation and angular momentum $5^-$. Combined with the fact that due to the high relative orbital angular momentum of the cluster motion the penetration of the clusters will be small, we then expect this to be a state with a relatively high energy expectation value. Therefore our argument that no low excited $5^-$ state in $O^{16}$ exists remains valid, even if the oscillator assumption no longer holds. This example again shows how the results which one obtains with oscillator cluster functions are also approximately valid for generalized cluster wave functions as long as the deviations from the oscillator cluster functions are not too large.

We wish now to indicate how the flexibility of generalized cluster functions can be fully exploited. Our development of these functions starting from the oscillator assumption was primarily for the purposes of a simple and useful first formulation. Having semi-quantitatively verified the validity of the physical concepts in the cluster approach, there is no need to stay with internal and relative motion functions of oscillator form. A *deuteron*-cluster for example is expected to be rather broad and diffuse; hence the internal function of this cluster is liable to be better represented by a superposition of Hulthen (exponential tail) functions than by a single Gaussian (see chap. VI sect. B). On the other hand an $\alpha$-cluster is rather tightly packed, so one or two Gaussians should suffice for its internal function. Further, if we wish to calculate cross-sections for nuclear scattering and reactions we cannot stick to bound state boundary conditions*. We can handle such problems by applying scattering boundary conditions to the relative motion functions (see chap. V). We can further modify generalized cluster functions to include the effects of hard-core forces in a theoretically consistent way.

There is also no reason to stay within a single cluster representation. We can build new complete sets by using a sufficient number of cluster functions of different cluster structure. This is mathematically unnecessary, for each cluster representation generates a complete set; but for purposes of physical insight and practical calculation it is more useful to employ an expansion system each of whose terms is more or less physically interpretable. For example, we know the energy is very

---

* So long as we are interested in just the energies of low-lying virtual levels then bound state boundary conditions are not a bad approximation, because the region where all the nucleons are strongly interacting is surrounded by a high Coulomb barrier and often an angular momentum barrier also, and the wave function must decay exponentially through these barriers, see for instance K. WILDER-MUTH and Y. TANG [41].

sensitive to the kind of cluster structure present; but it would be difficult to investigate the importance of simultaneously strong $a$-$d$ and $t$-$He^3$ cluster correlations in the states of $Li^6$ if we were to stick just to, say, the $t$-$He^3$ cluster representation. By using the sum of an $a$-$d$ cluster function and a $t$-$He^3$ cluster function we can affect the strength (i.e. how tightly the cluster nucleons are packed) of each of these clusters with just a few parameters (the internal widths). The amplitudes of the two cluster function can also be used as additional variational parameters, which we can roughly interpret as controlling the relative amount of each cluster structure present*. In general the addition of different cluster structures is particularly significant for conditions at the nuclear surface; this is of especial importance in nuclear fission and nuclear scattering and reaction problems where it is connected with the number of open and closed channels. By using mixed cluster representations we are able to control these effects with just a few variational quantities. This is to be compared with the many, almost uninterpretable parameters required if we remain with a single arbitrary complete set. Because we know at least their approximate physical significance, we can usually make a judicious choice of one or a few generalized cluster functions to find a good approximate solution of the $A$-nucleon Schrödinger equation in the frame of the Ritz variational principle.

### C. More detailed calculations of the matrix elements

The cluster description of nuclear states presented so far gives a theoretical picture, justified from first principles and capable of quantitative results, by which to understand the main features of nuclear behaviour. The price we pay is that by the introduction of collective cluster coordinates, the calculations can become quite difficult. In this section we shall show how by taking advantage of certain symmetries the calculations can be made practical, at least for lighter nuclei. We shall afterward introduce a useful mathematical representation for certain general considerations about collective properties. To discuss the calculations in detail we again consider the example of $Be^8$ with two unexcited $a$-clusters [21], L. PEARLSTEIN et al. [42]:

$$\psi(Be^8) = \mathscr{A}\,\phi$$
$$= \mathscr{A}\,\{\phi_0(a_1)\,\phi_0(a_2)\,\chi(R)\,Z(R_{CM})\,\xi(s,t)\}. \qquad (IV,7)$$

---

* Because the two cluster functions are not orthogonal we cannot interpret every parameter exactly. For instance, increasing the $a$-cluster internal width parameter strengthens mainly the $a$-cluster, but to a lesser extent the $t$-, $He^3$-, and $d$-clusters, too. This does not lessen the approximate significance of these parameters: taken together they control the nature of the cluster structure present. Hence they will be significant variational parameters in any calculation. The non-orthogonality causes no practical problem so long as the total wave function is kept normalized. The interpretation of the individual parameters becomes more definite the more we leave the oscillator assumption.

In the treatment of Be$^8$ we are concerned with calculating matrix elements

$$\langle \mathscr{A} \phi_{\text{final}} |O| \mathscr{A} \phi_{\text{initial}} \rangle .$$

As a first obvious simplification we may write using the Hermitian property of the antisymmetrizer $\mathscr{A}$ and the property $\mathscr{A}^2 = A!\mathscr{A}$:

$$\langle \mathscr{A} \phi_{\text{fin}} |O| \mathscr{A} \phi_{\text{in}} \rangle = A! \langle \mathscr{A} \phi_{\text{fin}} |O| \phi_{\text{in}} \rangle = A! \langle \phi_{\text{fin}} |O| \mathscr{A} \phi_{\text{in}} \rangle \quad \text{(IV,8)}$$

where $O$ must be an operator symmetric in all nucleon coordinates. Thus for such symmetric operators $O$ we need carry out the antisymmetrization only on the initial state or the final state, but not both.

Examples:

Expectation value of the potential energy;

$$O = V = \frac{1}{2} \sum_i \sum_{j \neq i} V_{ij} \quad ; \phi_{\text{in}} = \phi_{\text{fin}} . \quad \text{(IV,9)}$$

Expectation value of the kinetic energy in the center of mass system:

$$O = \bar{T} \equiv T - T_{\text{CM}} = \sum_i \frac{p_i^2}{2M} - \frac{P_{\text{CM}}^2}{2AM}$$

$$= \sum_i \frac{p_i^2}{2M} - \frac{1}{2AM} \left( \sum_i p_i \right)^2; \phi_{\text{in}} = \phi_{\text{fin}} \quad \text{(IV,9a)}$$

where $p_i = \frac{\hbar}{i} \nabla_i$.

Expectation value of the charge density at the position $r'$:

$$O = \sum_i e \delta(r_i - r') \cdot \frac{1}{2}(1 + \tau_{z_i}); \phi_{\text{in}} = \phi_{\text{fin}} \quad \text{(IV,10)}$$

where the factor $\frac{1}{2}(1 + \tau_{z_i})$ is 1 for proton states and 0 for neutron states.

Square of the normalization constant;

$$O = 1; \phi_{\text{in}} = \phi_{\text{fin}} . \quad \text{(IV,11)}$$

Electric quadrupole transition matrix elements, for instance

$$O = Q_e = \frac{1}{4} e \sqrt{\frac{15}{2\pi}} \sum_j (x_j + iy_j)^2 \frac{1}{2}(1 + \tau_{z_j}); \quad \phi_{\text{in}} \neq \phi_{\text{fin}} . \quad \text{(IV,12)}$$

In most of these matrix elements the center of mass motion is irrelevant. Hence the center of mass function $Z(R_{\text{CM}})$, which appears explicitly in cluster functions, may be split off and the matrix elements integrated over just the relative and internal coordinates. Unless otherwise noted the reader may assume this has been done.

The considered Be$^8$ wave functions without antisymmetrization explicitly written down with all coordinates have the following form

$$\phi(r_1, \ldots r_8) = \phi_0(\bar{r}_1, \ldots \bar{r}_4) \phi_0(\bar{r}_5, \ldots \bar{r}_8) \chi(R) \xi(s_1, \ldots t_8) Z(R_{\text{CM}}) \quad \text{(IV,13)}$$

where

$$\phi_0(\bar{r}_1, \ldots \bar{r}_4) = \exp\left\{-\frac{a}{2}\,(\bar{r}_1^2 + \bar{r}_2^2 + \bar{r}_3^2 + \bar{r}_4^2)\right\}$$

$$\chi(\boldsymbol{R}) = R^{2n} e^{-bR^2} Y_{lm}(\theta_R, \varphi_R)$$

$$\xi(s_1, \ldots t_8) =$$

$$= \binom{1}{0}_{s_1}\binom{1}{0}_{t_1}\binom{0}{1}_{s_2}\binom{1}{0}_{t_2}\binom{1}{0}_{s_3}\binom{0}{1}_{t_3}\binom{0}{1}_{s_4}\binom{0}{1}_{t_4}\binom{1}{0}_{s_5}\binom{1}{0}_{t_5}\binom{1}{0}_{s_6}\binom{0}{1}_{t_6}\binom{1}{0}_{s_7}\binom{1}{0}_{t_7}\binom{0}{1}_{s_8}\binom{0}{1}_{t_8}$$

$$Z(\boldsymbol{R}_{\mathrm{CM}}) = e^{-4cR_{\mathrm{CM}}^2}$$

and the cluster coordinates $\boldsymbol{R}, \bar{r}_1, \bar{r}_2, \bar{r}_3, \bar{r}_5, \bar{r}_6, \bar{r}_7$ are defined by equation (III,6) and (III,10). The dependent coordinates $\bar{r}_4$ and $\bar{r}_8$ are also defined there. $l$ is even since alpha particles are Bose particles. When we antisymmetrize we get an $S = 0$, $T = 0$ wave function. If $l$ would be odd, the wave function automatically would become 0 by antisymmetrization. To perform a definite calculation let us take for $O$ a potential energy operator consisting of a sum of two-particle interaction potentials.

$$V_{jk} = V_0 \exp\left\{-\beta(r_j - r_k)^2\right\}\{w(1 + P_{jk}^r)\} \qquad (\mathrm{IV},14)$$

where the operator $P_{jk}^r$ exchanges the space coordinates of the $j^{\mathrm{th}}$ and $k^{\mathrm{th}}$ nucleon. The two-body potential (IV,14) is equal to the two-body potential (III,14) if there we put $b = 0$. Due to the special symmetry of our assumed Be[8] wave function in regard to the spin and isospin dependence, this omitted part of the interaction potential (III,14) has the expectation value zero. For the case of antisymmetrized wave functions it is sometimes useful to write for the space-exchange operator in (IV,14):

$$P_{jk}^r = -P_{jk}^q P_{jk}^\tau \qquad (\mathrm{IV},15)$$

where $P_{jk}^q = \frac{1}{2}\,(1 + \boldsymbol{\sigma}_j \cdot \boldsymbol{\sigma}_k)$ and $P_{jk}^\tau = \frac{1}{2}\,(1 + \boldsymbol{\tau}_j \cdot \boldsymbol{\tau}_k)$ are the spin and isobaric spin (or charge) exchange operators. (IV,15) can be easily proved from the antisymmetrization of the total wave function:

$$P_{jk}^r P_{jk}^q P_{jk}^\tau \psi_{(\text{antisymmetrized})} = P_{jk}\psi_{(\text{antisymmetrized})} = -\psi_{(\text{antisymmetrized})}.$$

Then, multiplying from the left by $P_{jk}^r$ and using that $P_{jk}^r \cdot P_{jk}^r$ is equal 1:

$$P_{jk}^q P_{jk}^\tau \psi_{(\text{antisymmetrized})} = -P_{jk}^r \psi_{(\text{antisymmetrized})}. \qquad (\mathrm{IV},16)$$

The result (IV,16) shows that the representation (IV,15) is valid so long as we restrict ourselves to totally antisymmetrized wave functions.

Proceeding with the calculation of the potential energy expectation value, from (IV,8) we take the matrix element in the form $\langle \mathscr{A}\phi|V|\phi\rangle$ where $\phi$ is given by (IV,13). The number of terms in the matrix element is, due to $\mathscr{A}\phi$, quite large; but we shall now see how the symmetry of the wave function considerably reduces the number of terms that must be separately evaluated. The spin and isospin part of the wave function may be written more shortly:

$$\xi = (\uparrow\,\uparrow)_1\,(\downarrow\,\uparrow)_2\,(\uparrow\,\uparrow)_3\,(\downarrow\,\uparrow)_4\,(\uparrow\,\uparrow)_5\,(\downarrow\,\uparrow)_6\,(\uparrow\,\uparrow)_7\,(\downarrow\,\uparrow)_8 \qquad (\mathrm{IV},17)$$

where $\uparrow_1$ means particle one has spin up, $\downarrow_2$ means particle two has isospin down etc. Now if, as in our example, the operator $O$ contains no spin or isospin exchange operators, it cannot change the spin isospin function of the initial state*. Hence the only terms in the antisymmetrized final state wave function which contribute to the expectation value of $O$ are the terms which have the same identical spin and isobaric spin configuration as the unsymmetrized initial state; if they do not have the same configuration as the initial state then they are orthogonal to this state and their contribution to the matrix element of $O$ vanishes.

In our example, where $\phi_{\text{in}} = \phi_{\text{fin}} = \phi$, inspection of the spin isospin part (IV,17) of the wave function $\phi$ shows that only a limited number of permutations have this spin isospin configuration unaltered. Of the terms which contribute, there is first the term in which no particles are exchanged, called the direct term, and the term where the two clusters are totally exchanged, which we call the four particle exchange term. But because our Be$^8$ wave function is symmetric in the two $\alpha$-clusters we observe these two terms are identical. Going on, there are four possible nonvanishing one particle exchange terms, $\{1 \leftrightarrow 5\}$, $\{2 \leftrightarrow 6\}$ etc. These are equal to each other and also equal to the three particle exchange terms, of which there are also four. This is again due to the symmetry character of our Be$^8$ wave functions. These exchange terms have a negative sign because an odd number of pair exchanges are performed. Finally for the two particle exchange terms the sign of the permutation is positive. Examples of non-vanishing two particle-exchange terms are:

$$\begin{Bmatrix} 1 \leftrightarrow 5 \\ \text{and} \\ 2 \leftrightarrow 6 \end{Bmatrix} ; \quad \begin{Bmatrix} 2 \leftrightarrow 6 \\ \text{and} \\ 4 \leftrightarrow 8 \end{Bmatrix} .$$

Of these there are six contributing terms, which again, due to the symmetry character of our Be$^8$ wave functions, are all equal to each other.

The result of the expectation value of $O = \frac{1}{2} \sum_{jk} V_{jk}$ is thus reduced to

$$N^2 \langle O \rangle = 8! \int [2\phi^*(\boldsymbol{r}_1, \ldots \boldsymbol{r}_8) - 8\phi^*(\boldsymbol{r}_5 \boldsymbol{r}_2 \boldsymbol{r}_3 \boldsymbol{r}_4 \boldsymbol{r}_1 \boldsymbol{r}_6 \boldsymbol{r}_7 \boldsymbol{r}_8) +$$
$$+ 6\phi^*(\boldsymbol{r}_5 \boldsymbol{r}_6 \boldsymbol{r}_3 \boldsymbol{r}_4 \boldsymbol{r}_1 \boldsymbol{r}_2 \boldsymbol{r}_7 \boldsymbol{r}_8)] \left\{ \frac{1}{2} \sum V_{jk} \right\} \phi(\boldsymbol{r}_1 \ldots \boldsymbol{r}_8) \, d\tau \qquad \text{(IV,18)}$$

where $d\tau$ represents integration over a suitable set of coordinates. The wave functions are unnormalized, and the expectation value of $O$ must be corrected later for normalization, which is the purpose of the $N^2$ factor. The function $\phi$ here denotes only the spatial part of the unsymmetrized cluster function (IV,13); the inner product of the spin isospin functions for contributing terms is unity, hence no spin isospin dependence appears in this expression.

---

* If in the operator $O$ spin (or isobaric spin) exchange operators appear, then in the matrix element only those terms in the antisymmetrized final state contribute which have the same spin and isobaric spin configurations as $P^\sigma_{jk}\phi_{\text{in}}$ (or $P^\tau_{jk}\phi_{\text{in}}$).

We now wish to turn to another problem concerned not so much with the practical evaluation of matrix elements but rather with the mathematical formulation of more general arguments to be used later[*]. These arguments require a specific designation of the cluster coordinates of an antisymmetrized cluster function for the purpose of performing mathematical manipulations directly on the cluster coordinates. In a wave function describing physically indistinguishable nucleons the problem arises how to unambiguously specify a set of cluster coordinates mathematically without giving preferential labels to some of the nucleons.

The problem is overcome by the method of parameter coordinates, in which the cluster coordinates can be introduced completely distinct from the nucleon coordinates, [21]. To introduce the method consider at first the nucleons 1 and 2 in the same spin and isospin states and located at the points $r'_1$ and $r'_2$. If the nucleons were distinguishable with 1 at $r'_1$ and 2 at $r'_2$, the space part of the wave function would be

$$\phi \propto \delta(r_1 - r'_1)\,\delta(r_2 - r'_2)\;\text{[**]}$$

where $r_1$ and $r_2$ are the nucleon coordinates and $r'_1$ and $r'_2$ are the parameter coordinates. Since the nucleons obey the Pauli principle we must antisymmetrize, obtaining

$$\phi \propto \{\delta(r_1 - r'_1)\,\delta(r_2 - r'_2) - \delta(r_2 - r'_1)\,\delta(r_1 - r'_2)\}\,.$$

This wave function does not tell which nucleon is at $r'_1$ or $r'_2$; this means that the parameter coordinates which describe the position of the nucleons are completely separated from the nucleon coordinates. Therefore we can now introduce the center of mass and relative coordinates of the two nucleons in the parameter coordinates:

$$R' = \frac{1}{2}(r'_1 + r'_2);\quad r' = r'_1 - r'_2\,.$$

With this we obtain

$$\phi \propto \left\{\delta\left(r_1 - \left(R' + \frac{r'}{2}\right)\right)\delta\left(r_2 - \left(R' - \frac{r'}{2}\right)\right)\right.$$
$$\left. - \delta\left(r_2 - \left(R' + \frac{r'}{2}\right)\right)\delta\left(r_1 - \left(R' - \frac{r'}{2}\right)\right)\right\}. \tag{IV,19}$$

If we now give $R'$ and $r'$ fixed values we have in no way indicated any nucleon to be preferentially fixed to one of the occupied points and our originally problem is solved.

Let us now apply the method to our Be[8] wave function (IV,7) and (IV,13)

$$\psi(r_1\ldots r_8) = \mathscr{A}\{\phi_0(a_1)\,\phi_0(a_2)\,\chi(R)\,Z(R_{CM})\}$$
$$= \mathscr{A}\int \phi_0(a'_1)\,\phi_0(a'_2)\,\chi(R')\,Z(R'_{CM})\,[\delta(r_1 - r'_1) \quad \text{(IV,20)}$$
$$\delta(r_2 - r'_2)\cdots\delta(r_8 - r'_8)]\,\mathrm{d}r'_1\ldots\mathrm{d}r'_8$$

---

[*] See for example the discussion of rotational states, sec.E.
[**] $\delta(r_1 - r'_1)$ means $\delta(x_1 - x'_1)\,\delta(y_1 - y'_1)\,\delta(z_1 - z'_1)$ and similarly for $\delta(r_2 - r'_2)$.

where the delta function $\delta(\underset{\sim}{r_1} - \underset{\sim}{r_1'})$ and so on, now includes the spin and isobaric spin coordinates of the nucleon as well as its space coordinates, and the integration includes summation over the discrete parameter spin and isobaric spin coordinates:

$$\delta(\underset{\sim}{r_1} - \underset{\sim}{r_1'}) \equiv \delta(r_1 - r_1')\, \delta_{s_1 s_1'} \delta_{t_1 t_1'} \tag{IV,21}$$

$$\int d\tau' \equiv \int d\underset{\sim}{r_1'}\, d\underset{\sim}{r_2'} \ldots d\underset{\sim}{r_8'} = \sum_{s_1'} \sum_{t_1'} \sum_{s_2'} \cdots \sum_{t_8'} \int dr_1'\, dr_2' \ldots dr_8'.$$

The wave function is unchanged by this manipulation, as can be seen by carrying out the integration. The antisymmetrization is carried out on the nucleon coordinates $r_1, s_1, t_1, \ldots, r_8, s_8, t_8$, not on the primed parameter coordinates which are distinct from the nucleon coordinates. We can introduce for the primed (parameter) coordinates now the internal coordinates and the center of mass coordinates of the clusters, which is just a change of the dummy variables of integration.

Using the definitions (III,6) and (III,10), but with all coordinates there primed to denote that the transformations are carried out on the parameter coordinates and not the nucleon coordinates, we have

$$\psi(\underset{\sim}{r_1} \ldots \underset{\sim}{r_8}) = \int \phi_0(\underset{\sim}{a_1'})\, \phi_0(\underset{\sim}{a_2'})\, \chi(\underset{\sim}{R'})\, Z(\underset{\sim}{R_{\mathrm{CM}}'}) \times$$

$$\times \mathscr{A}\left[ \delta\left( r_1 - \left( R_{\mathrm{CM}}' + \frac{1}{2} R' + \bar{r}_1' \right) \right) \delta_{s_1 s_1'} \delta_{t_1 t_1'} \times \right. \tag{IV,22}$$

$$\left. \times\, \delta\left( r_2 - \left( R_{\mathrm{CM}}' + \frac{1}{2} R' + \bar{r}_2' \right) \right) \delta_{s_2 s_2'} \delta_{t_2 t_2'} \ldots \text{etc} \ldots \right] d\tau'.$$

In this integral representation of the Be$^8$ wave function the cluster coordinates are now completely distinct from the nucleon coordinates and may be directly manipulated without treating any nucleon as distinguishable.

While the integral form (IV,22) avoids none of the practical labor of calculation, it can lead to alternative expressions giving considerable physical insight into nuclear behaviour. To see how such alternative expressions can arise let us again evaluate a potential energy matrix element $\langle \psi | U | \psi \rangle$, this time using the parameter method. Since our example is only for purposes of insight, we shall simplify the two nucleon potential to infinitely short-ranged forces

$$U = \frac{1}{2} \sum U_{ik} = \frac{1}{2} \sum (- U_0 \delta(r_i - r_k)) \tag{IV,23}$$

and set the internal and relative motion widths of our Be$^8$ cluster function to a common value denoted by $a$ (oscillator assumption). Substituting the parameter coordinate integral representation for $\psi$, eq. (IV,20), into the matrix element for the potential, (IV,23), we proceed as in the earlier calculations: first removing the antisymmetrization operator from the right hand side, then using orthogonality of the spin isospin functions to remove vanishing terms, and finally collecting terms that are equivalent because of the symmetry of the cluster function.

We then obtain

$$N^2 \langle \psi | U | \psi \rangle = 8! \int \int \int d\tau \, d\tau' \, d\tau'' \{ \phi(r_1'' \ldots r_8'') \times$$

$$\times [2\delta(r_1 - r_1'') \ldots \delta(r_8 - r_8'') -$$

$$- 8 \, \delta(r_5 - r_1'') \, \delta(r_2 - r_2'') \ldots \delta(r_1 - r_5'') \ldots \delta(r_8 - r_8'') + \quad \text{(IV,24)}$$

$$+ 6\delta(r_5 - r_1'') \, \delta(r_6 - r_2'') \ldots \delta(r_8 - r_8'')] \} \cdot \left\{ \frac{1}{2} \sum U_{ik} \right\} \times$$

$$\times \{ \phi(r_1' \ldots r_8') \, [\delta(r_1 - r_1') \ldots \delta(r_8 - r_8')] \}$$

where

$$\phi(r_1'' \ldots r_8'') = \phi(a_1'') \, \phi(a_2'') \, \chi(R'') \, Z(R_{\text{CM}}'')$$

and

$$\int d\tau' = \int dr_1' \ldots dr_8' = \int d R_{\text{CM}}'' \, dR' \, d\bar{r}_1' \, d\bar{r}_2' \, d\bar{r}_3' \, d\bar{r}_5' \, d\bar{r}_6' \, d\bar{r}_7' . \quad \text{(IV,25)}$$

Excepting the fact that we have here retained the center of mass function and have used a modified potential, this result is identical to (IV,18), as a simple integration over the parameters coordinates shows. Alternatively (IV,24) can be derived from (IV,18) by substituting in parameter coordinates for the spatial coordinates there.

We now substitute primed and double primed cluster parameter coordinates in (IV,24) for the primed and double primed parameter coordinates respectively, using the usual transformations (III,6) and (III,10).

Eq. (IV,24) is now integrated over all nucleon coordinates and all cluster parameter coordinates excepting $d^3 R''$ and $d^3 R'$*. Then we obtain, see [21]

$$\langle O \rangle = - V_0 \, 3 \cdot 8 \cdot 8! \, \frac{B_1 + B_2 - 2B_3 - B_4 - 4B_5 + B_6 + 4B_7}{N^2} \quad \text{(IV,26)}$$

with

$$B_1 = \frac{\pi^9}{a^9} \int R'^{2(n+1)} \, e^{-2aR'^2} |Y_{l,m}(\theta', \varphi')|^2 \, dR' \, d\Omega'$$

$$B_2 = 2^3 \left( \frac{\pi^6}{3a^6} \right)^{\frac{3}{2}} \int R'^{2(n+1)} \, e^{-\frac{4}{3}aR'^2} |Y_{l,m}(\theta', \varphi')|^2 \, dR' \, d\Omega'$$

$$B_3 = 2^6 \left( \frac{\pi^5}{6a^5} \right)^{\frac{3}{2}} \int R''^{n+2} R'^{n+2} \, e^{-\frac{5}{3}a(R''^2 - R'' \cdot R' + R'^2)} Y_{l,m}^*(\theta'', \varphi'') \times$$

$$\times Y_{l,m}(\theta', \varphi') \, dR'' \, dR' \, d\Omega'' \, d\Omega'$$

$$B_4 = 2^3 \left( \frac{\pi^5}{a^5} \right)^{\frac{3}{2}} \int R''^{n+2} R'^{n+2} \, e^{-a(3R''^2 - 2R'' \cdot R' + 3R'^2)} Y_{l,m}^*(\theta'', \varphi'') \times$$

$$\times Y_{l,m}(\theta', \varphi') \, dR'' \, dR' \, d\Omega'' \, d\Omega'$$

---

* The irrelevant center of mass integrals cancel with corresponding factors in the normalization constant $N^2$. Thus if we had used plane wave functions $\exp\{i k \cdot R_{\text{CM}}\}$ we would still obtain a finite result despite the $\delta$-function normalization of such functions. Alternatively for calculational purposes we can replace the plane wave function by $\exp\{- 4 \, c R_{\text{CM}}^2\}$ or any such center of mass wave function normalized to unity. The results are unchanged and no infinities arise.

$$B_5 = 2^3 \left(\frac{\pi^5}{a^5}\right)^{\frac{3}{2}} \int R''^{n+2} R'^{n+2} e^{-a(3R''^2 - 4R'' \cdot R' + 4R'^2)} Y^*_{l,m}(\theta'', \varphi'') \times$$
$$\times Y_{l,m}(\theta', \varphi')\, dR''\, dR'\, d\Omega''\, d\Omega'$$

$$B_6 = 2^3 \left(\frac{\pi^5}{2a^5}\right)^{\frac{3}{2}} \int R''^{n+2} R'^{n+2} e^{-2a(R''^2 + R'^2)} Y^*_{l,m}(\theta'', \varphi'') \times$$
$$\times Y_{l,m}(\theta', \varphi')\, dR''\, dR'\, d\Omega''\, d\Omega'$$

$$B_7 = 2^6 \left(\frac{\pi^5}{6a^5}\right)^{\frac{3}{2}} \int R''^{n+2} R'^{n+2} e^{-a\left(\frac{8}{3}R''^2 + 2R'^2\right)} Y^*_{l,m}(\theta'', \varphi'') \times$$
$$\times Y_{l,m}(\theta', \varphi')\, dR''\, dR'\, d\Omega''\, d\Omega'$$

$$d\Omega'' = \sin\theta''\, d\theta''\, d\varphi''$$
$$d\Omega' = \sin\theta'\, d\theta'\, d\varphi'. \tag{IV,27}$$

In deriving these, the integrations over the delta functions can be carried out most conveniently in cartesian coordinates, and afterwards one introduces other coordinates, as for instance spherical coordinates.

For the normalization one finds

$$N^2 = 2 \cdot 8! \cdot \left(\frac{2\pi}{a}\right)^3 (B_1 - 4B_3 + 3B_6). \tag{IV,28}$$

Integrals of the form (IV,27) appear typically in nuclear structure and reaction problems, even if we use Gaussian two-nucleon interactions and generalized cluster wave functions.

In this special example the remaining integrals can also be carried out explicitly. For completeness we give the results, see [21].

$$B_1 = \sqrt{\frac{\pi}{2a}}\, \frac{\pi^9}{a^{n+10}}\, \frac{(2n+1)!!}{2^{2n+3}}$$

$$B_2 = \sqrt{\frac{\pi}{2a}}\, \frac{\pi^9}{a^{n+10}}\, \frac{(2n+1)!!}{2^{2n+3}} \left(\frac{3}{4}\right)^n$$

$$B_3 = \sqrt{\frac{\pi}{2a}}\, \frac{\pi^9}{a^{n+10}} \frac{3^{n-1}}{2^{4n+3}}\, \frac{[(n+l+1)!!]^2}{(2l+1)!!} \times$$
$$\times\ _2F_1\left(-\frac{1}{2}(n-l),\ -\frac{1}{2}(n-l);\ l+\frac{3}{2};\ \frac{1}{4}\right)$$

$$B_4 = \sqrt{\frac{\pi}{2a}}\, \frac{\pi^9}{a^{n+10}} \frac{3^{n-l}}{2^{4n+3}}\, \frac{[(n+l+1)!!]^2}{(2l+1)!!} \times$$
$$\times\ _2F_1\left(-\frac{1}{2}(n-l),\ -\frac{1}{2}(n-l);\ l+\frac{3}{2};\ \frac{1}{9}\right)$$

$$B_5 = \sqrt{\frac{\pi}{2a}}\, \frac{\pi^9}{a^{n+10}} \frac{3^{1/2(n-l)}}{2^{3n+3}}\, \frac{[(n+l+1)!!]^2}{(2l+1)!!} \times$$
$$\times\ _2F_1\left(-\frac{1}{2}(n-l),\ -\frac{1}{2}(n-l);\ l+\frac{3}{2};\ \frac{1}{3}\right)$$

$$B_6 = \sqrt{\frac{\pi}{2a}}\, \frac{\pi^9}{a^{n+10}} \frac{1}{2^{2n+3}}\, [(n+1)!!]^2\, \delta_{l0}$$

$$B_7 = \sqrt{\frac{\pi}{2a}}\, \frac{\pi^9}{a^{n+10}} \frac{1}{2^{3n+3}}\, [(n+1)!!]^2\, \delta_{l0}. \tag{IV,29}$$

In (IV,29) $k!! = 1 \cdot 3 \ldots k$ for odd $k$ and $2 \cdot 4 \ldots k$ for even $k$; ${}_2F_1(a, b; c; d)$ is the hypergeometric function of the indices 2, 1 and the arguments $a, b, c, d$, see [21] page 456. Further, for obtaining the expressions (IV,29) from the expressions (IV,27) the expansion

$$e^{-\lambda \boldsymbol{R}'' \cdot \boldsymbol{R}'} = e^{\lambda R'' R' \cos \gamma}$$

$$= \sqrt{4\pi [2l + 1]} \sum_{l=0}^{\infty} (-1)^l \sqrt{\frac{\pi}{2\lambda R'' R'}} \, I_{l+\frac{1}{2}}(\lambda R'' R') \, Y_{l,0}(\gamma)$$

$$= 4\pi \sum_{l=0}^{\infty} \sum_{m=-l}^{l} (-1)^l \sqrt{\frac{\pi}{2\lambda R'' R'}} \, I_{l+\frac{1}{2}}(\lambda R'' R') \times$$

$$\times Y_{lm}(\theta_R'', \varphi_R'') \, Y_{lm}(\theta_R', \varphi_R') \tag{IV,30}$$

was used. $I_{l+\frac{1}{2}}(\lambda R'' R')$ are the modified cylinder functions with half-integer indices of the imaginary argument $i\lambda R' R''$.

To obtain the final expressions (IV,29) it is not necessary to introduce the parameter coordinates explicitly. One can introduce in the right hand unsymmetrized wave function directly the cluster variables, and will obtain the final expressions (IV,29) much faster than by the just sketched procedure. But by doing this one will miss the intermediate expressions (IV,27). Such intermediate expressions will be important for later general discussions.

## D. The influence of cluster overlapping

In chapter III section $D$ we discussed, mainly in the frame of the oscillator model, the influence of the Pauli principle upon cluster wave functions. We now resume this discussion, but without the restriction to the oscillator assumption. We shall be especially interested in how the effect of the Pauli principle increases with increasing mutual penetration of the clusters. We shall pursue the question with the aid of two detailed examples, H. HACKENBROICH et al. [43]. In the first we shall see how the Pauli principle usually causes clusters to differ greatly from the corresponding ordinary composite particles as the clusters increasingly penetrate each other. In the second however, we shall see that a cluster still remains energetically favored even in nuclear matter, where it is completely surrounded and penetrated by a Fermi sea of nucleons.

For our first example we consider again a two $\alpha$-cluster function. To begin our discussion we need a measure of how the Pauli principle alters this cluster function from the simple particle picture, such as would be represented here by the $\alpha$ dumb-bell model. Let $\phi$ be a two cluster function antisymmetrized only within each cluster and let $P'$ denote just those permutations exchanging nucleons between the clusters but not within the individual clusters. We define

$$OV \equiv \left| \frac{\sum_{P'} (-1)^{P'} \langle P'\phi \mid \phi \rangle}{\langle \mathscr{A}'\phi \mid \phi \rangle} \right|^2 \tag{IV,35}$$

where

$$\mathscr{A}'\phi = \phi + \sum_{P'} (-1)^{P'} P' \phi$$

where the sum goes over sufficient permutations of the type $P'$ to effect a complete antisymmetrization of the cluster function*. We refer to $OV$ as the degree of overlapping between the clusters. When the clusters are overlapped very little ($OV \ll 1$), in the calculation of matrix elements usually the antisymmetrization need only be carried out within the individual clusters. Only in the latter case does it make sense to treat the clusters as ordinary particles. Therefore the degree of overlapping, $OV$ is a quantitative measure of the influence of the Pauli principle between the clusters.

Let us now make some general arguments as to how the degree of overlapping depends upon the mutual spatial penetration of the two clusters. Suppose first that the two clusters do not penetrate each other, that is, the most probable separation distance of the cluster centroids is much larger than the sum of the "radii" of the two clusters. In this case $P'\phi$ as a function of the permuted nucleon coordinates will be small in a region where $\phi$ as a function of the nucleon coordinates is large, and vice versa. Consequently $OV$ will be much less than unity. Suppose on the other hand the most probable separation distance of the clusters is extremely small compared to the sum of the radii of the clusters. By the uncertainty principle the kinetic motion of the two clusters about one another then becomes very large. In this semi-classical limit situation we again expect the influence of the Pauli principle between the clusters to be small. To see this is in fact the case we consider the cluster function in the momentum representation. We see that the internal momentum spread within a cluster is much smaller than the momentum spread of the relative motion**. Hence $P'\phi$ as a function of the permuted nucleon momenta is small where $\phi$ as a function of the nucleon momenta is large, and vice versa. Hence the degree of overlapping $OV$ will again be

---

* Note the sum does not include all such permutations $P'$ but only a sufficient number that the antisymmetrization is complete. For instance in a *triton-deuteron* cluster state of He⁵ we have

$$\phi = \mathscr{A}_t \phi_0(1\,2\,3)\,\mathscr{A}_d \phi_0(4\,5)\,\chi(1\,2\,3 - 4\,5)$$

where if $P_I$ is the identity permutation and $P_{ij}$ a pair exchange, $\mathscr{A}_t$ and $\mathscr{A}_d$ are given by

$$\mathscr{A}_t = P_I - P_{12} - P_{13} - P_{23} + P_{13}P_{12} + P_{23}P_{12}$$
$$\mathscr{A}_d = P_I - P_{45}$$

and $\mathscr{A}'$ is represented by

$$\mathscr{A}' = P_I + \sum_{P'} (-1)^{P'} P'$$

$$= P_I + (-P_{14} - P_{15} - P_{24} - P_{25} - P_{34} - P_{35} + P_{14}P_{25} + P_{14}P_{35} + P_{24}P_{35})$$

where we define $\mathscr{A}'$ by $\mathscr{A} \equiv \mathscr{A}'\mathscr{A}_t\mathscr{A}_d$ for the triton deuteron case.

** This is obvious for the width-parameters in the momentum representation are proportional to the reciprocals of the respective width-parameters in the position representation.

small despite the considerable spatial penetration*. Between these two extremes $OV$ will become large and we expect the behaviour of the antisymmetrized cluster function to deviate considerably from the ordinary particle picture.

Let us verify these general arguments specifically for our two $\alpha$-cluster example. The two $\alpha$-cluster function is from (IV,13)

$$\psi(\underset{\sim}{r}_1 \ldots \underset{\sim}{r}_8) = \mathscr{A} \, N_{lm} \left( e^{-\frac{a}{2} \sum\limits_{i=1}^{4} \vec{r}_i^2} \right) \left( e^{-\frac{a}{2} \sum\limits_{i=5}^{8} \vec{r}_i^2} \right) R^n e^{-bR^2} Y_{lm}(\theta_R, \varphi_R) \cdot \xi(s_1 \ldots t_8)$$

(IV,36)

where $N_{lm}$ is the normalization constant. The ratio of the relative motion width to the internal width $b/a$ is a measure of the spatial penetration of the clusters. For $b = a$ the wave function (IV,36) coincides with the $Be^8$ oscillator cluster function, and the clusters penetrate each other strongly. For $b/a \ll 1$ the clusters are relatively far apart.

We will now compare the charge densities resulting from the wave function (IV,36) with and without antisymmetrization as a function of $b/a$. From (IV,10) we have for the expectation value of the charge density

$$\varrho_e(r) = 4e \sum_{s_i} \sum_{t_i} \int \psi^*(\underset{\sim}{r}_1 \ldots \underset{\sim}{r}_8) \, \psi(\underset{\sim}{r}_1 \ldots \underset{\sim}{r}_8) \, \mathrm{d}r_2 \ldots \mathrm{d}r_8 ; r = r_1 \quad \text{(IV,36a)}$$

where the sum goes over all spin and isospin coordinates. The results of the calculations** are shown in Fig. 10–12.

For small penetration of the clusters, $(b/a) = \frac{1}{4}$, the antisymmetriza-tion practically does not influence the charge distribution. For large penetration of the clusters, $(b/a) = 1$ and $(b/a) = 4$, the charge distribu-tion is altered considerably by the antisymmetrization. The change is always in the direction of reducing the charge density for small $r$ values and increasing it at larger $r$ values. This is understandable for when the clusters begin to overlap the Pauli principle forces nucleons of the same spin and isospin to repel each other. For $b/a$ much greater than 5 in our example the large distributions again become essentially the same for both the unsymmetrized and antisymmetrized wave function. The behaviour of the charge density is thus just what we had expected from our general arguments.

Next we compare the electric quadrupole transition probabilities, $|Q(E_2)|^2$ from the $2^+$ level to the ground state as a function of a cluster penetration $b/a$ again for both the antisymmetrized and unsymmetrized wave function (IV,36). From (IV,12) we have for $|Q(E_2)|^2$:

$$|Q(E_2)|^2 = \frac{15 e^2}{32 \pi} \left| \langle \psi_{\text{fin}} | \sum_j (x_j + i y_j)^2 \frac{1}{2} (1 + \tau_{zj}) | \psi_{\text{in}} \rangle \right|^2 \quad \text{(IV,37)}$$

---

* This second extreme case will not be realized in nature because the large relative oscillation would destroy the internal cluster structure.
** We use $n = 4$, $m = l = 0$ in (IV,36) for the calculation.

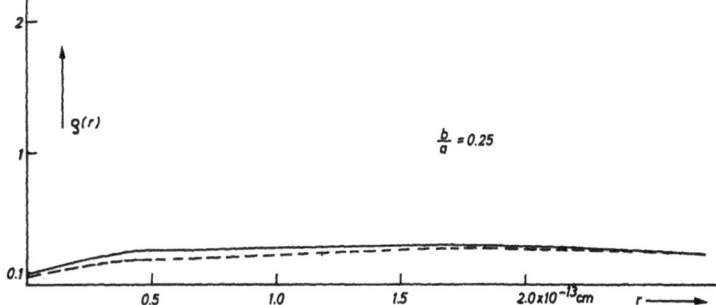

Fig. 10. Charge distribution of Be⁸ cluster function with clusters rather separated; solid line: unsymmetrized, dotted line: antisymmetrized

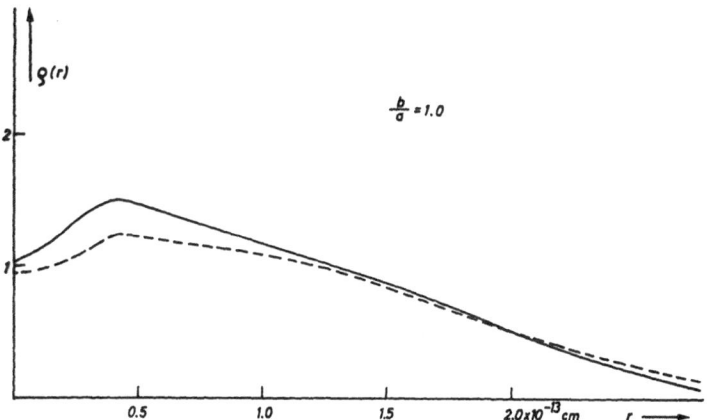

Fig. 11. Charge distribution of Be⁸ cluster function with clusters strongly penetrating one another; solid line: unsymmetrized, dotted line: antisymmetrized

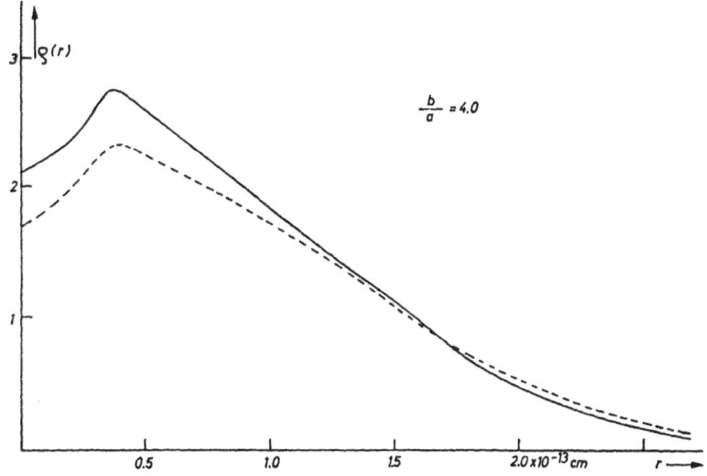

Fig. 12. Charge distribution of Be⁸ cluster function with clusters strongly penetrating one another; solid line: unsymmetrized, dotted line: antisymmetrized

where $\psi_{\text{fin}}$ and $\psi_{\text{in}}$ denote the final ground state and initial $2^+$ state respectively. In Fig. 13 the ratio $|Q(E_2)|^2/|Q^0(E_2)|^2$ is plotted vs. $b/a$, where $|Q^0(E_2)|^2$ is the value of (IV,37) for the oscillator cluster model value $(b/a) = 1$. The calculation is made assuming the deformation of the nucleus to be the same in both the initial and final state, i.e. only $l$ was changed from 0 to 2 in (IV,36) to obtain the wave function of the $(2^+)$ state.

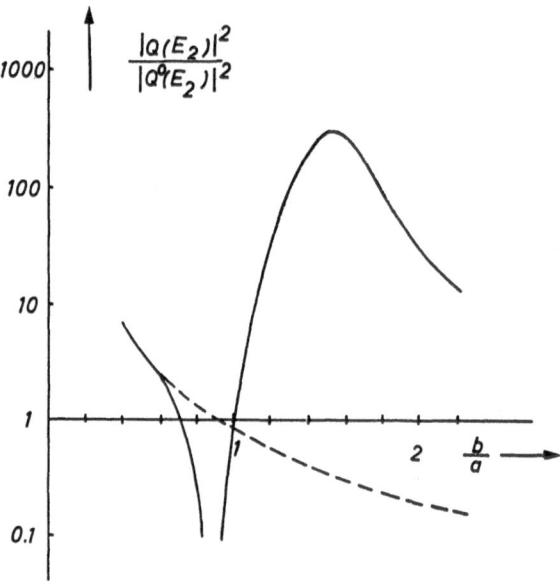

Fig. 13. Dependence of electric quadrupole transition probability upon cluster separation distance; dotted curve: calculated with unsymmetrized wave functions, solid curve: calculated with antisymmetrized wave functions

The most remarkable feature of Fig. 13 is the behaviour of the curve near $b/a$ about 0.9. For this particular value of $b/a$ the electrical quadrupole transition probability approaches zero due to an interference effect. This effect is connected with the Pauli principle, as can be seen from the dashed curve where the antisymmetrization of the wave functions is not taken into account*. The dashed curve has a very smooth behaviour as a function of $b/a$ and does not approach zero for any finite $b/a$ value. For small $b/a$ values the solid curve approaches the dashed curve, as it must, because as the average distance of the *alpha* clusters becomes very large the antisymmetrization becomes less and less important. The same is true if $b/a$ becomes very large because, as already mentioned, due to the large relative momentum of the clusters the overlapping of the alpha-particle cluster wave functions again becomes very small.

The calculations with the two $\alpha$-cluster wave function discussed above have borne out our original arguments that, excepting when the spatial

---

    * A more detailed investigation shows that the one and three nucleon exchange terms between the clusters are mainly responsible for this interference effect [43].

penetration of the clusters is slight or when their relative momentum is great, the degree of cluster overlapping is large and the behaviour of the antisymmetrized cluster function can be completely different from that of the unsymmetrized function. We repeat that these arguments are general and not restricted to the above typical example.

Therefore in speaking about clusters one has to remember that the nucleons are indistinguishable. No individual nucleon can be said to belong to a particular cluster, whereas this cluster may well always contain the same number of nucleons. One may imagine a continuous exchange of nucleons going on between the clusters (from the time required for a nucleon to traverse the nucleus we infer the exchange time $\approx 10^{-23}$ sec). This nucleon exchange decreases very rapidly with decreasing spatial penetration of the clusters and vanishes when the clusters are completely separated. The adequate expression of this is the antisymmetrization operator $\mathscr{A}$.

Although cluster overlapping destroys the naive "anschauliche" concept of intranuclear structures appreciably, one important property remains which fundamentally influences all nuclear structure and reaction problems. If a cluster with a tightly bound closed shell configuration is surrounded by many other nucleons, then even if the degree of overlapping is large this cluster correlation usually remains energetically favoured over any kind of state in which the closed shell cluster is broken up. We want to justify this conclusion in some detail.

To understand this continued energetical preference of cluster correlations we consider a simple one-dimensional model where the essential features creating the effect can be seen especially clearly. The model consists of a Fermi sea of nucleons into which we introduce a cluster of a finite number of nucleons. The motion of the uncorrelated nucleons in the Fermi sea we describe by single particle wave functions, eigenfunctions to an infinite potential well of diameter $L$. The internal motion of the nucleons in the cluster we describe also by single particle eigenfunctions, to an infinite potential well, but of diameter $l \ll L$. We have not used an oscillator potential to build up the cluster because we wish to keep the calculation simple; for the same reason we assume the total momentum of the cluster relative to the Fermi sea is zero. By these assumptions a zero point oscillation of the cluster center of mass remains, but this has no essential influence on our considerations.

The unnormalized wave function of our total system is now described by the following expression

$$\psi = \mathscr{A}'(\phi_0 \phi_1) \tag{IV,38}$$

where $\phi_0$ represents the nucleons in the Fermi sea and $\phi_1$ the nucleons in the cluster. $\phi_0$ and $\phi_1$ are both antisymmetrized functions. $\mathscr{A}'$ describes the antisymmetrization between the Fermi sea and the cluster (see footnote page 53). The normalized wave function $\phi_0$ has the form

$$\phi_0 = \frac{1}{\sqrt{A_0!}} \left(\frac{2}{L}\right)^{\frac{A_0}{2}} \mathscr{A}_0 \prod_{n=1}^{\frac{A_0}{4}} \sin\left(\frac{2\pi n x_n}{L}\right) \xi_n; \quad -\frac{L}{2} \leq x_n \leq \frac{L}{2}. \tag{IV,39}$$

$A_0$ is the number of the nucleons in the Fermi sea. The $\xi_n$ are spin-isospin functions. Therefore to every $n$ value belong four different spin-isospin states if we assume that all space states are filled up with four nucleons.

For the normalized wave function $\phi_1$ one obtains analogous to (IV,39)

$$\phi_1 = \frac{1}{\sqrt{A_1!}} \left(\frac{2}{l}\right)^{\frac{A_1}{2}} \mathscr{A}_1 \prod_{m=1}^{\frac{A_1}{4}} \sin\left(\frac{2\pi m x_m}{l}\right) \xi_m; \quad -\frac{l}{2} \leq x_m \leq \frac{l}{2} \qquad \text{(IV,40)}$$

with $A_1$ the number of the nucleons in the cluster.

For simplicity we have assumed that all single particle wave functions in $\phi_0$ and $\phi_1$ vanish at the origin. That means we use only single particle wave functions with odd parity. This restriction does not influence the generality of our considerations because the following discussion can be applied in exactly the same way to the even parity states. In (IV,39) and (IV,40) we have assumed that in the ground state of the Fermi sea and of the cluster the lowest single particle levels of the potential wells allowed by the Pauli principle are filled up.

One recognizes immediately that single particle wave functions in the cluster are changed completely by the antisymmetrization operator $\mathscr{A}'$ in (IV,38) if their wave number $k_m = 2\pi m/l$ is smaller or a little larger than the wave number $K_F = 2\pi A_0/4L$ of the Fermi limit. On the other hand those single particle wave functions of the cluster with $k_m$ appreciably greater than $K_F$ are changed only very little.

To see this in more detail we expand the single particle wave functions of the cluster in single particle wave functions of the Fermi sea. Defining $K_n = \frac{2\pi n}{L}$ we have

$$\phi_1(k_m) = \sqrt{\frac{2}{l}} \sin(k_m x) \xi_m$$
$$= \sum_{n=1}^{\infty} a_{K_n}(k_m) \sqrt{\frac{2}{L}} \sin(K_n x) \xi_m \qquad \text{(IV,41)}$$

where after some transformations we find

$$a_{K_n}(k_m) \equiv \int_{-\frac{l}{2}}^{\frac{l}{2}} \sqrt{\frac{2}{l}} \sin(k_m x) \sqrt{\frac{2}{L}} \sin(K_n x) \, dx$$

$$= \begin{cases} \dfrac{4}{\sqrt{Ll}} (-1)^m \dfrac{k_m \sin\left(\frac{K_n l}{2}\right)}{K_n^2 - k_m^2} & ; k_m \neq K_n \\[3mm] \sqrt{\dfrac{l}{L}} & ; k_m = K_n. \end{cases} \qquad \text{(IV,42)}$$

We consider now $a_K(k)$ in the two limit cases $k \gg K_F$ and $k < K_F$. In the first case we assume further that the wave number distance $\Delta k = 2\pi/l$ of two successive levels in the cluster is much smaller than $k - K_F$, that is, many levels of the cluster lie above the Fermi sea.

In the first case $a_K(k)$ has its maximum for $K \approx k$ and $a_K(k) = \sqrt{l/L}$. The wave number width of the considered level is of the order of $\Delta k$ because already for

$$K \approx k \pm \frac{\pi}{l} \tag{IV,43}$$

$|a_K(k)|^2$ decreases by a factor 0.4, which means such a cluster level is practically not influenced by the presence of the Fermi sea.

In the second case $k < K_F$ the situation is different. Here in the presence of the Fermi sea $a_K(k)$ must be zero for $k < K_F$, due to the Pauli principle. That is, $k_m < K_F$ implies the amplitudes of (IV,42) are drastically altered to

$$\bar{a}_K(k_m) = \begin{cases} (-1)^m N_m \dfrac{\sin\left(\dfrac{Kl}{2}\right)}{K^2 - k_m^2} & ; K > K_F \\ 0 & ; K \leq K_F \end{cases} \tag{IV,44}$$

The normalization constant $N_m$ is defined by the condition

$$\sum_{K=K_F}^{\infty} |\bar{a}_K(k_m)|^2 = 1 .$$

We assume now further that $K_F - k_m$ is much larger than $\Delta k$. That means we consider a cluster nucleon which is bedded deeply into the Fermi sea. For the altered single particle wave function of this cluster level we obtain then

$$\bar{\phi}_1(k_m) = \sum_{K=K_F}^{\infty} \bar{a}_k(k_m) \sqrt{\frac{2}{L}} \sin Kx$$

$$\approx N_m \int_{K_F}^{\infty} a_K(k_m) \sqrt{\frac{2}{L}} \sin Kx \, dK \tag{IV,45}$$

$$\sim (-1)^m \left\{ -\frac{\cos\left(\frac{l}{2} + x\right) K_F}{\left(\frac{l}{2} + x\right)} + \frac{\cos\left(\frac{l}{2} - x\right) K_F}{\left(\frac{l}{2} - x\right)} \right.$$

where to perform the approximate integration in (IV,45) we used $K_F - k_m$ much larger than $\Delta k$. It can be shown easily that (IV,45) does not diverge for $x \to \pm \frac{l}{2}$.

One sees from (IV,45) that due to the presence of the Fermi sea, so long as we consider energy levels of the cluster which are bedded deeply into the Fermi sea, the single particle wave functions $\phi_1(k_m)$ are much more smeared out over the whole space than the original single particle wave functions (IV,41) which vanish for $|x| \geq l/2$. This means that the original correlations between the cluster nucleons are reduced very much by the antisymmetrization, particularly for the low energy cluster nucleons (see our earlier remarks, chap. III sect. A).

We use these results now for the discussion of the energy expectation value of the total system consisting of the Fermi sea and the cluster.

The normalized wave function of this system we can write as follows

$$\psi = \sqrt{\frac{A_0! A_1!}{A!}} \, \mathscr{A}' \, (\phi_0 \cdot \phi_1) \tag{IV,46}$$

where $A = A_0 + A_1$.

In (IV,46) $\phi_0$, which represents the nucleons in the Fermi sea, is the same wave function as $\phi_0$ in (IV,38). $\phi_1$ consists again of an anti-symmetrized product of single particle wave functions as in $\phi_1$ in (IV,38). But the single particle wave functions are now defined by

$$\phi_1(k_m) = \sum_{K=K_F}^{\infty} \bar{a}_K(k_m) \sqrt{\frac{l}{2}} \sin(Kx) \tag{IV,46a}$$

where $\bar{a}_K(k_m)$ is defined in (IV,42).

The $\phi_1(k_m)$ are no more orthogonal to each other, as were the original single particle wave functions of the cluster. But they are now all ortho-gonal to the single particle wave functions of the Fermi sea. Further the $\phi_1(k_m)$ of the highest cluster levels, $k_m \gg K_F$, are also approximately orthogonal to all other $\phi_1(k_m)$. In $\phi_1$ a normalization factor is contained which normalizes $\phi_1$ to one. This factor depends on the mutual overlap-ping of all single particle wave functions. $\mathscr{A}'$ describes again the anti-symmetrization between the Fermi sea and the cluster. Therefore a factor $\sqrt{A_0! A_1!/A!}$ appears in front of the wave function (IV,46), necessary to normalize it.

We consider first the expectation value of the kinetic energy of our system. With (IV,46) we obtain

$$\langle \psi | T | \psi \rangle = \left\langle \phi_0 \left| \sum_{n}^{A_0} \left( -\frac{\hbar^2}{2M} \nabla_n^2 \right) \right| \phi_0 \right\rangle + \left\langle \phi_1 \left| \sum_{m}^{A_1} \left( -\frac{\hbar^2}{2M} \nabla_m^2 \right) \right| \phi_1 \right\rangle \tag{IV,47}$$

where in order to keep the notation transparent we mean by the sums

$$\sum_{n}^{A_0} = \sum_{n=1}^{A_0} \quad \text{and} \quad \sum_{m}^{A_1} = \sum_{m=A_0+1}^{A}.$$

The first term in (IV,47) is the kinetic energy of the Fermi sea. It is

$$E_{\text{KIN}}^F \equiv \left\langle \phi_0 \left| \sum_{n=1}^{A_0} -\frac{\hbar^2}{2M} \nabla_n^2 \right| \phi_0 \right\rangle$$

$$= \sum_{n=1}^{A_0/4} 4 \int \frac{\hbar^2}{2M} K_n^2 \left( \frac{2}{L} \right) \sin^2(K_n x_n) \, dx_n$$

$$= \sum_{K_n=0}^{K_F} \frac{2\hbar^2}{M} K_n^2 \approx \frac{2\hbar^2}{M} \frac{2\pi}{L} \int_0^{K_F} K^2 dK$$

$$= \frac{1}{6} \frac{\pi^2 \hbar^2}{M} \frac{A_0^3}{L^2}. \tag{IV,48}$$

The second term in (IV,47) is the kinetic energy of the nucleons in the cluster, where the cluster wave function $\phi_1$ in influenced by the presence of the Fermi sea in the just described manner. No overlapping term between the cluster wave function $\bar{\phi}_1$ and the wave function $\phi_0$ of the Fermi sea appears in (IV,47). This is due to the fact that the kinetic energy operator is a single particle operator.

We consider now the potential energy of our system. If we assume two-particle forces $V_{ik}$ as interaction forces between the nucleons then we have for the potential energy expectation value of our system

$$E_{\text{POT}} = \left\langle \psi \left| \frac{1}{2} \sum_i^A \sum_{k \neq i}^A V_{ik} \right| \psi \right\rangle.$$

If we use for our "ansatz" (IV,46) then the potential energy splits up into the following terms

$$E_{\text{POT}} = \left\{ \left\langle \phi_0 \left| \sum_i^{A_0} \sum_{k \neq i}^{A_0} \frac{1}{2} V_{ik} \right| \phi_0 \right\rangle + \left\langle \mathscr{A}' \phi_0 \bar{\phi}_1 \left| \sum_i^{A_0} \sum_k^{A_1} V_{ik} \right| \phi_0 \bar{\phi}_1 \right\rangle + \right.$$

$$\left. + \left\langle \bar{\phi}_1 \left| \sum_i^{A_1} \sum_{k \neq i}^{A_1} \frac{1}{2} V_{ik} \right| \bar{\phi}_1 \right\rangle \right\}. \qquad (\text{IV},49)$$

For the derivation of (IV,49) it is essential that $\phi_0$ and $\bar{\phi}_1$ are normalized to one and their single particle wave functions are orthogonal to each other. Therefore, any exchange term between the Fermi sea and the cluster vanishes in the first and the third term of (IV,49).

The first term of (IV,49) represents the potential energy of the Fermi sea and the third term the potential energy of the cluster. The second term is the interaction energy of the cluster with the Fermi sea.

For these calculations we assume a simplified form for our two body forces $V_{ik}$. To simulate the short range character of the nuclear forces we choose attractive $\delta$-forces of the form $V_{ik} = -V_0 \delta(x_i - x_k)$ for the two-particle interactions, as is often done in nuclear physics.

If we introduce

$$\varrho_0(x_1, x_2) = 4 \sum_{n=1}^{A_0/4} \frac{2}{L} \sin\left(\frac{2\pi n}{L} x_1\right) \sin\left(\frac{2\pi n}{L} x_2\right)$$

$$\varrho_0(x_1) \equiv \varrho_0(x_1, x_1) \qquad (\text{IV},50)$$

then we obtain for the potential energy of the Fermi sea, HEISENBERG [44],

$$E_{\text{POT}}^{\text{F}} = \frac{1}{2} \int dx_1 dx_2 V_{12} \left\{ \varrho_0(x)_1 \varrho_0(x_2) - \frac{1}{4} |\varrho_0(x_1, x_2)|^2 \right\}$$

$$= -\frac{1}{2} V_0 \int \frac{3}{4} \varrho_0^2(x_1) \, dx_1 \approx -\frac{3}{8} V_0 \frac{A_0^2}{L}. \qquad (\text{IV},51)$$

Inspection of (IV,50) shows that $\varrho_0(x_1)$ is identical to the number density of nucleons in the Fermi sea. In (IV,51) it was assumed that we have many nucleons in the Fermi sea, therefore $\varrho_0(x_1)$ becomes approximately $A_0/L$.

By introducing

$$\varrho_1(x) = A_1 \sum_{s_i} \sum_{t_i} \int |\bar{\phi}_1(x_1 \ldots x_{A_1})|^2 \, dx_2 \ldots dx_{A_1} \qquad (\text{IV},52)$$

where $x = x_1$, we obtain similarly to (IV,51) for the interaction energy between the cluster and the Fermi sea

$$E_{\text{POT}}^{\text{INT}} = -\frac{3}{4} V_0 \int \varrho_0(x)\, \varrho_1(x)\, \mathrm{d}x \cong -\frac{3}{4} V_0 \int \varrho_1(x)\, \frac{A_0}{L}\, \mathrm{d}x$$

$$= -\frac{3}{4} V_0 \frac{A_1 A_0}{L}. \tag{IV,53}$$

The factor $\frac{3}{4}$ in (IV,53) appears due to cancellations from the exchange terms in the antisymmetrized functions $\phi$ and $\bar\phi_1$ analogous to the factor $\frac{3}{4}$ in (IV,51). We see that $E_{\text{POT}}^{\text{INT}}$ does not depend on the special form of the cluster wave function $\bar\phi_1$ as long as the number of nucleons in the Fermi sea, $A_0$, is sufficiently large.

To evaluate $E_{\text{POT}}^{\text{C}}$ we cannot proceed as before for $E_{\text{POT}}^{\text{F}}$ because the single particle cluster wave functions (IV,46a) are not all orthogonal to each other.

We introduce therefore the probability to meet a cluster nucleon* in the volume element $\mathrm{d}x_r$ and at the same time another cluster nucleon in the volume element $\mathrm{d}x_s$.

This probability is given by

$$w_1(x_r, x_s)\, \mathrm{d}x_r\, \mathrm{d}x_s \tag{IV,54}$$

$$= \frac{1}{2} A_1 (A_1 - 1) \sum_{s_1} \cdots \sum_{t_{A_1}} \{ \int |\bar\phi_1(x_1 \ldots x_{A_1})|^2\, \mathrm{d}x_3 \ldots \mathrm{d}x_{A_1} \}\, \mathrm{d}x_r\, \mathrm{d}x_s$$

with

$$x_r = x_1; \quad x_s = x_2.$$

With (IV,54) we obtain for the potential energy of the cluster

$$E_{\text{POT}}^{\text{C}} = \left\langle \bar\phi_1 \left| \frac{1}{2} \sum_i^{A_1} \sum_k^{A_1} V_{ik} \right| \bar\phi_1 \right\rangle = \frac{1}{2} \int V_{rs}\, w_1(x_r x_s)\, \mathrm{d}x_r\, \mathrm{d}x_s$$

$$= -\frac{1}{2} V_0 \int_{-L/_2}^{L/_2} w_1(x_r, x_r)\, \mathrm{d}x_r. \tag{IV,55}$$

Eq. (IV,55) leads to quite significant results. The potential energy of the cluster depends on the integral of the two nucleon probability function for the cluster nucleons, $w_1(x, x)$. The integral has its least value if $w_1(x, x)$ is constant over the volume of the Fermi sea. It has its greatest value if $w_1(x, x)$ is sharply peaked with a large maximum value somewhere in the space. But this means that the density of cluster nucleons, $\varrho_1(x)$ must also be sharply peaked with a large maximum value. Hence the stronger the cluster correlation (here determined by the cluster well

---

* This probability is not usually a physically measurable one, for when the cluster is greatly overlapped by surrounding nucleons it is not possible to say definitely which nucleons are participating in the cluster correlation at a given moment. Nevertheless $w_1(x_r, x_s)$ is well defined and enters in the calculation of the potential energy.

width-parameter $l$) the greater the contribution of the cluster to the potential energy*.

Opposing this tendency of the potential energy to cluster the nucleons strongly together are the kinetic energy, the Pauli principle, and the hard core of the nuclear forces (we have not included hard-core forces explicitly in the calculation; clearly they do not affect the overall nature of eq. (IV,55). But as they are the principle source of saturation we discuss their role here). These three effects limit, often drastically, how tightly the cluster nucleons may be packed together. As we decrease the cluster well width $l$, the kinetic energy of the cluster nucleons increases due to the uncertainty principle. The Pauli principle forbids nucleons in the same spin and isospin state to come close to one another. No nucleons may come so close that they penetrate the hard-core regions. Since both the Pauli principle and the hard-core forces serve to limit the available volume in which the nucleons may move, they show up here mainly as an increased contribution to the kinetic energy. The nucleons will pack together as closely as possible until the rise in kinetic energy, and possibly repulsive core potential energy, offset any further gain in potential energy from the short-range attractive forces. This important conclusion shows that the clusters remain energetically favoured even when surrounded and greatly overlapped by many other nucleons.

We consider now the break-up energy of the cluster. By break-up energy we mean the energy required to remove a nucleon from the cluster and place it in one of the unoccupied states of the Fermi sea, $K > K_F$. Unfortunately, this break-up energy is very difficult to calculate for realistic cases. From our experience with light nuclei calculations we have found that the break-up energy of a cluster is usually less but of the same order of magnitude as the break-up energy of the corresponding free particle. In order to make a calculation of the break-up energy of the cluster in the Fermi sea we will make the rather special assumption that the highest occupied levels of the cluster lie well above the level of the Fermi sea. We emphasize that this assumption is not necessary for our previous conclusion about the energetical preference of the clusters; eqs. (IV,47−55) are derived without it. With this assumption we can split up the cluster function $\tilde{\phi}_1$ as follows:

$$\tilde{\phi}_1 = \sqrt{\frac{A_2! A_3!}{A_1!}}\ \mathscr{A}'(\phi \cdot \chi)$$

$$A_1 = A_2 + A_3 .\tag{IV,56}$$

$\phi$ consists of the antisymmetrized product of the single particle wave functions $\tilde{\phi}_1(k_m)$ of all cluster levels except the highest ones. $A_2$ is the number of these lower cluster levels. $\chi$ consists of the antisymmetrized product

---

* In fact it is obvious from eq. (IV,51) that since the potential energy of a Fermi sea depends on the integral of the square of the nucleon density, then we could anticipate that its potential energy is increased by introducing cluster correlations, thereby increasing the density in some parts and decreasing it in others.

of the single particle wave functions of the $A_1 - A_2 = A_3$ highest cluster levels. $\phi$ and $\chi$ are normalized to one. $\mathscr{A}'$ describes the antisymmetrization between $\phi$ and $\chi$. To keep the notation simple we always understand that before antisymmetrization the first $A_0$ nucleons are in the Fermi sea, the nucleons from $A_0 + 1$ to $A_0 + A_2$ constitute the group $A_2$ of lower energy cluster nucleons, and the nucleons $A_0 + A_2 + 1$ to $A$ constitute the group $A_3$ of high energy cluster nucleons.

Due to our discussion above $\chi$ can be written in good approximation as

$$\chi = \frac{1}{\sqrt{A_3!}} \left(\frac{2}{l}\right)^{A_3/2} \mathscr{A}_3 \prod_m^{A_3/4} \sin(k_m\, x_m)\, \varepsilon_m . \tag{IV,57}$$

The single particle wave functions in (IV,57) are orthogonal to each other. Further they are approximately orthogonal to the single particle wave functions of the nucleons in the Fermi sea and less rigorously to the single particle wave functions in $\phi$ *. That means, this approximation for $\chi$ is very nearly orthogonal to $\phi_0$ and $\phi$.

With this we obtain for the kinetic energy of the cluster wave function $\bar{\phi}_1$ (second term in (IV,47))

$$E_{KIN}^C = \left\langle \bar{\phi}_1 \left| \sum_n^{A_1} -\frac{\hbar^2}{2M}\, \nabla_n^2 \right| \bar{\phi}_1 \right\rangle$$

$$= \left\langle \phi \left| \sum_n^{A_2} -\frac{\hbar^2}{2M}\, \nabla_n^2 \right| \phi \right\rangle + \left\langle \chi \left| \sum_m^{A_3} -\frac{\hbar^2}{2M}\, \nabla_m^2 \right| \chi \right\rangle . \tag{IV,58}$$

For the second term in (IV,58) using (IV,57) we can write in good approximation

$$\left\langle \chi \left| \sum_m^{A_3} -\frac{\hbar^2}{2M}\, \nabla_n^2 \right| \chi \right\rangle = 4\, \frac{\hbar^2}{2M} \sum_m^{A_3/4} k_m^2 . \tag{IV,59}$$

If we now break up the cluster by removing a nucleon from one of the highest occupied cluster single particle levels to an unoccupied level of the Fermi sea, then to a good approximation the kinetic energy change will be described by the variation of the term (IV,59) only. But as long as one neglects rearrangement effects, see for instance P. MITTEL-STAEDT [45], this kinetic energy variation is the same as in the case where one considers the breaking up of the cluster without a surrounding Fermi sea**.

We consider now the potential energy of our system, eq. (IV,49). The contribution to the potential energy from the first two terms $E_{POT}^F + E_{POT}^{INT}$ of (IV,49) is unaltered by break-up, because as we pointed out $E_{POT}^{INT}$ does not depend on the density arrangement of the cluster. This means the change in the total potential energy due to break-up comes just from the change in the cluster potential energy $E_{POT}^C$ itself.

---

* We repeat once more the single particle wave functions in $\phi$ are not all orthogonal to each other.

** Except that now no restriction due to the Pauli principle exists as to which state outside of the cluster one can place the cluster nucleon in. To this point we return shortly.

For more detailed discussion of the cluster potential energy we again employ our ansatz for $\bar{\phi}_1$ in (IV,56). With this we obtain analogously to (IV,49) through (IV,53)

$$E_{POT}^C = \left\langle \phi \left| \sum_i^{A_2} \sum_k^{A_2} -\frac{1}{2} V_0 \delta(x_i - x_k) \right| \phi \right\rangle + \tag{IV,60}$$
$$-\frac{3}{4} V_0 \int \varrho_2(x)\, \varrho_3(x)\, dx - \frac{3}{8} V_0 \int |\varrho_3(x)|^2\, dx\,.$$

The first term and the third term in (IV,60) describe the interaction among the low energy (which we shall call "inner") cluster nucleons and among the high energy ("outer") nucleons respectively. The second term represents the interaction between the outer and inner nucleons.

Analogously to (IV,55) the potential energy of the inner cluster nucleons can be written as

$$E_{POT}^{inner} \equiv \left\langle \phi \left| \sum_i^{A_2} \sum_k^{A_2} -\frac{1}{2} V_0 \delta(x_i - x_k) \right| \phi \right\rangle$$
$$= -\frac{1}{2} V_0 \int w_2(x_r, x_r)\, dx_r \tag{IV,61}$$

where $w_2(x_r, x_s)\, dx_r\, dx_s$ is the "probability" to meet at the same time one inner cluster nucleon in the volume element $dx_r$ and another inner cluster nucleon in the volume element $dx_s$ (see however footnote on p. 62). From (IV,60) and (IV,61) we see once more that the potential energy of the cluster will become large if the nucleons in the cluster are packed as tightly as possible because (IV,60) and (IV,61) will become large if

$$\varrho_2(x) = \int w_2(x, x_s)\, dx_s \quad \text{and} \quad \varrho_3(x)$$

are large in some local region of the Fermi well.

We now consider the change of the potential energy if we break up the cluster by removing a nucleon of the highest occupied cluster levels to an unoccupied state of the Fermi sea. We see from our formulas (IV,60) and (IV,61) that the only terms which will change are the interaction among just the outer cluster nucleons themselves and the interaction between the outer cluster nucleons and the inner cluster nucleons. Again analogously to the corresponding kinetic energy change, the interaction energy change of the outer cluster nucleons is in good approximation the same as if no Fermi sea were present as long as we neglect rearrangement effects [45]. The positive energy change (decreasing binding energy) is

$$\Delta E_{POT}^{outer} = \frac{3}{16} V_0 \int |\varrho_3(x)|^2\, dx\,. \tag{IV,61a}$$

The energy change of the interaction energy between the inner and outer cluster nucleons will be also positive and is given by

$$\Delta E_{POT}^{inner\text{-}outer} = \frac{3}{16} V_0 \int \varrho_2(x)\, \varrho_3(x)\, dx\,. \tag{IV,62}$$

This energy change is smaller than the corresponding energy change of the free cluster because, as we pointed out earlier, the nucleon density $\varrho_2(x)$ of the inner nucleons is spread out over the whole space more in the case where a surrounding Fermi sea is present than where no surrounding Fermi sea is present. If one also takes into account the exclusion principle between the inner cluster nucleons themselves then $\varrho_2(x)$ is smeared out even more, because the new single particle wave functions of the inner cluster nucleons, eq. (IV,46a), are no longer orthogonal.

This decrease of the positive energy change by going from the free cluster case to the case where the cluster is surrounded by other nucleons is partly compensated by the following effects. First the negative kinetic energy change of the cluster will be larger for the free cluster than for a surrounded cluster because in the former case no states outside of the cluster are already occupied by nucleons of the Fermi sea. Second, rearrangement energy effects, which always increase the binding energy of the system, are again due to the Pauli principle usually larger in the former case than in the latter. Therefore the energy needed to break up the cluster will be of the same order of magnitude whether the cluster is surrounded by other nucleons or not*. So we see in this calculable, though special, case the trend found in light nuclei for the value of the break-up energy remains valid.

We have now to ask if the arguments for the energetical favouring of clusters, discussed following (IV,55), are restricted to our special example or do they hold in general. The essential point was that the attractive $\delta$-forces want to pack the nucleons as tightly as possible. While we do not have $\delta$-forces in reality, we do have average attractive forces of finite but very short range. Hence (IV,55) clearly reflects the principle trend of the real forces. A second point is that we have not included the hard-core forces explicitly. However suppose we start with a Fermi sea of completely uncorrelated nucleons representing nuclear matter. The average density of nuclear matter is such that there is only a small probability for nucleons to come so close together that they penetrate the hard-core region, L. GOMES et al. [46], A. DE SHALIT and V. WEISSKOPF [47]. Hence, aside from determining the equilibrium density, hard-core effects are not very significant for the wave function of the Fermi sea. But we have seen in (IV,51) that the potential energy of the Fermi sea depends on the integral of the square of the density. Then we can anticipate immediately that the potential energy is increased by introducing cluster correlations, thereby increasing the density in some parts of the Fermi sea and decreasing it in others. Eventually the density in the cluster can be made sufficiently great that hard-core effects will be significant, as well as the rise in kinetic energy and Pauli principle effects. But before we have reached this point we have already

---

* In this example we have restricted ourselves to the case where the cluster as a whole is at rest. Our considerations are certainly valid also if the cluster oscillates relative to the Fermi sea. For such an oscillation increases the average kinetic energy of every cluster nucleon and therefore enlarges the energy difference between them and the nucleons of the Fermi sea.

shown the energetical favouring for the existence of clusters. Hence the hard-core forces do not prevent the presence of clusters; rather they serve primarily only to help limit the final strength of the cluster correlations. Hence our use of simple attractive $\delta$-forces is justified.

Now let us consider the generality of our model, a cluster in a Fermi sea. The first point is that the important result (IV,55) is not dependent on the form of the well used to build up the cluster nor on the fact that the model is not three dimensional. Although the calculation is considerably more complicated, the same result is obtained if the cluster is constructed with an oscillator well. Nor does the Fermi sea approximation introduce any essential restriction. If we were to narrow the width of the Fermi sea and use a large but realistic number of nucleons, the same overall behaviour would recur. In fact if the larger and smaller cluster are described in the frame of the oscillator cluster model the general results are unchanged, as explicit but lengthy calculation shows. In particular the potential energy contribution of the cluster remains intimately related to the density of the nucleons in the cluster.

Finally we discuss what general inferences may be drawn regarding the most energetically favoured cluster configuration and the related cluster break-up energy. We begin with a plausible suggestion which we then immediately refine. After antisymmetrization with the Fermi sea the original cluster density is reduced considerably since the cluster nucleons (particularly the "inner" nucleons) become smeared out over the Fermi sea; it is plausible that this reduced density is maximally peaked if the original cluster before antisymmetrization with the Fermi sea was as densely packed as possible. We can draw an immediate conclusion. The original cluster before antisymmetrization with the Fermi sea corresponds to a free cluster, and it is well known that free clusters are exceptionally tightly packed when they are of closed shell configuration, i.e. clusters having only one or few neutrons and protons outside a closed (magic number) shell. If our plausibility argument is correct, a closed shell cluster in the Fermi sea will be energetically favoured over a non-closed shell cluster because even after antisymmetrization a closed shell cluster will be more densely packed than a non-closed shell cluster. Now let us refine the argument. If we expand an arbitrary antisymmetrized cluster function in oscillator shell model functions, the "inner" (low energy) nucleons will occupy completely filled oscillator shells. Nothing more can be done to correlate these inner nucleons further without breaking up the shells. On the other hand the outer nucleons occupy unfilled shells and so can be placed in a superposition of states which packs them tightly. Now we refer back to our calculation of the break-up energy of a cluster in the Fermi sea. Under the special assumption made there (that they lie well above the Fermi sea), the single particle levels of the outer nucleons are negligibly affected by antisymmetrization with the Fermi sea. Hence they are nearly identical to those of a free cluster.

Hence the outer nucleons of the cluster in the Fermi sea will be most tightly packed if the cluster is a magic number cluster. As just pointed out, since the outer nucleons are the only ones which can be significantly packed anyway, then closed shell clusters are more energetically favoured than non-closed shell clusters. Suppose we now relax our special assumption so that the outer nucleon levels need not lie far from the surface of the Fermi sea. It still remains that the outer nucleons are the ones least affected by antisymmetrization, and therefore it remains plausible that in a closed shell cluster the nucleons are packed more closely than in a non-closed shell cluster even when surrounded by other nucleons.

A further refinement gives some idea of the break-up energy of the cluster. In a shell model expansion of the antisymmetrized cluster function it is well known that the attractive nuclear forces act most strongly between nucleons in the same shell. Then the energy required to remove the last nucleon from the cluster will be greatest if the last shell is filled. However, if the last shell is filled or almost filled, the outer nucleons provide a significant part of the energy binding the last nucleon to the cluster. Since the outer nucleons are the least smeared out by the antisymmetrization, we expect for a closed shell cluster that the break-up energy to remove the last nucleon will not be too greatly reduced from that for the corresponding free cluster.

Hence we conclude generally that closed shell clusters usually remain energetically preferred even when highly overlapped by many surrounding nucleons, and that their break-up energy is usually less but of the same order of magnitude as their break-up energy when free.

We will now shortly apply our results to some specific examples. In the case of light nuclei one can explicitly show quantitatively, as we have often discussed already, that closed shell or nearly closed shell clusters, such as *triton*-clusters, $\alpha$-clusters, $O^{16}$-clusters, etc. in their ground states remain energetically favoured compared with other clusters, even if they are surrounded by other nucleons. *Deuteron*-clusters usually are even more tightly bound when they are bedded in other nucleons than when they are free particles (see chapter III section E).

In the case of heavy clusters such explicit calculations are not yet possible, and we have to rely on the results of our general considerations. As an example of heavy cluster behaviour, let us qualitatively examine the effects of cluster overlapping on a heavy closed shell cluster within a larger nucleus. For simplicity we consider a cluster having 50 neutrons. As far as cluster overlapping with these neutrons is concerned we can neglect the presence of the protons. Let us assume that our heavy nucleus has a total of 132 neutrons, so that the other cluster, having 82 neutrons, is also a closed shell cluster. These particular magic number clusters are of interest in connection with asymmetric fission. As usual, to determine what remains of a cluster correlation after antisymmetrization we will consider our example simultaneously in several representations: in the oscillator shell model, in the oscillator cluster model, and with generalized cluster functions.

In the oscillator shell model the energetically lowest state allowed by the Pauli principle for 132 neutrons has the first six oscillator shells filled and the seventh partially occupied by the last 20 neutrons, see App. B. This state has 540 oscillator quanta of energy. If this oscillator shell model function is now considered in conjunction with the nuclear Hamiltonian, the spin-orbit part of the nuclear forces strongly favours the outside nucleons to have $l$ and $s$ parallel with $l$ as large as possible. Hence 14 of the 20 outside neutrons will occupy the 14 possible single particle levels of the $7i$ subshell in which $l$ and $s$ are parallel. These 14 neutrons along with the 112 neutrons filling the first six oscillator shells comprise the nuclear closed shell configuration of 126 neutrons, M. GOEPPERT-MAYER and H. JENSEN [48]. The remaining 6 neutrons occupy other subshells of the $7^{th}$ oscillator shell.

In the oscillator cluster model we construct the internal functions of the clusters with single particle functions of an oscillator well*. Therefore due to the Pauli principle a cluster of 50 neutrons must have 130 oscillator quanta of internal energy. If we take into account the nuclear spin-orbit forces, the ten outside neutrons in the unfilled 5$^{th}$ oscillator shell will occupy the 10 levels of the $7g$ subshell in which $l$ and $s$ are alligned, thereby completing the nuclear 50 neutron closed shell. Similarly a cluster of 82 neutrons will have 270 oscillator quanta of internal energy, and the twelve neutrons in the unfilled 6$^{th}$ oscillator shell will occupy the 12 levels with $l$ and $s$ parallel of the $7h$ subshell [48]. The lowest relative oscillation of the two clusters must have at least 140 oscillator quanta of energy; for we have seen in the oscillator shell model that a wave function for 132 neutrons of less than 540 oscillator quanta vanishes under antisymmetrization. If we expand the oscillator cluster function in oscillator shell model functions, then, analogously to the Fermi sea example, under antisymmetrization any terms containing two nucleons in the same single particle level vanish.

Hence our oscillator cluster function corresponds to an antisymmetrized oscillator shell model function where the first six oscillator shells are filled and the last twenty nucleons are in some superposition of single particle levels of the 7$^{th}$ oscillator shell. Due to our magic number cluster assumption we expect this superposition will be such as to pack these last twenty nucleons as tightly as possible within the restriction of the Pauli principle. In the oscillator shell model these correlations between the last 20 nucleons are all that remains after antisymmetrization of our initial unsymmetrized cluster function. However, despite the fact that the Pauli principle drastically reduces the cluster correlations present in the unsymmetrized oscillator cluster function, the remaining correlations will energetically favour such a state over any other state in which the last 20 nucleons are less correlated.

If we now leave the oscillator assumption and use generalized cluster functions, then the energetical preference for the 50 and 82 neutron magic number cluster representation will increase. To see this, consider an expansion of the generalized cluster function in oscillator shell model

---

* We use an oscillator well again to construct the internal function, instead of a square well as in the cluster plus Fermi sea example, in order to facilitate comparison with the oscillator shell model.

functions. In this expansion will appear terms in which the 6 lowest oscillator shells are not completely filled and some higher oscillator levels are occupied. All the nucleons in unfilled oscillator shells can now be correlated so as to increase the nucleon density within the limits of the Pauli principle and by the general form of (IV,55) the potential energy will increase*. The break-up energy of the clusters is difficult to estimate. If we consider a reference frame in which the oscillator well of the 82 neutron cluster is at rest (analogous to the Fermi sea in our previous example) and expand the 50 neutron cluster function, which now has all the relative motion in addition to its internal energy, in states of the 82 neutron cluster oscillator well, then the most energetic neutrons of the 50 neutron cluster lie only one oscillator shell above the last occupied oscillator shell of the 82 neutron cluster. Consequently the simplifications of our special break-up energy calculation for the cluster in a Fermi sea do not obtain. But from our general remarks we expect the break-up energy to be somewhat smaller than that of the corresponding free cluster. It approaches the break-up energy of the free clusters more and more as we increase the relative motion of the two clusters.

If we had instead considered a 50 proton closed shell cluster within a larger nucleus, our considerations would have gone similarly. The main difference would be that in heavy nuclei the break-up energy of a 50 proton cluster is even larger than that of a 50 neutron cluster, because it is surrounded by less protons than the 50 neutron cluster by neutrons. We shall use these observations later when we take up the fission process in detail.

## E. Collective states

A great deal of insight into collective states can be obtained in the framework of cluster representations using the method of parameter coordinates (see chap. IV sect. C). In this section we will discuss particularly the rotational states of even-even nuclei, A. Bohr [49], A. Bohr and B. Mottelson [50]. By rotational states is usually meant a series of levels whose energies can be fit by an expression of the form:

$$E_l = A + Bl(l+1)$$

where $l$ is the (approximate) orbital angular momentum of the level and $A$ and $B$ are empirical constants almost independent of $l$. In the usual approach to such states the energy difference between the levels is supposed to arise mainly from differences in rotational kinetic energy, D. Inglis [51]. However we shall see that if antisymmetrization is properly taken into account, this energy difference may have nothing to do with rotational kinetic energy at all; it can arise purely from differences

---

* We might also note here that the nucleon density will be slightly increased near the surface of the clusters due to the spin-orbit forces, which tend to favour large nucleon orbital angular momentum values when $l$ and $s$ are parallel.

in the potential energy. In general the energy splitting of so-called rotational levels comes from both the kinetic and potential energies. We shall find it convenient therefore to generalize the definition of rotational states somewhat on the basis of their structure rather than their spectra.

Let us begin our investigation with $Be^8$ as a specific example. The rigid rotator or $\alpha$-particle dumb-bell model is typical of attempts to explain the energies of the three lowest states of $Be^8$ as rotational kinetic energy differences, J. WHEELER [51], J. PERRING and T. SKYRME [53]. In this approach we write the wave function for these levels as

$$\psi_{\rm rot}(Be^8) = (\mathscr{A}\,\phi_0(\underset{\sim}{a_1}))\,(\mathscr{A}\,\phi_0(\underset{\sim}{a_2}))\,f(R)\,Y_{lm}(\theta_{\rm R},\,\varphi_{\rm R})\,. \qquad (IV,63)$$

where $\phi_0(\underset{\sim}{a_1})$ and $\phi_0(\underset{\sim}{a_2})$ are the internal functions of the $\alpha$-particles and $\boldsymbol{R} = \boldsymbol{R_1} - \widetilde{\boldsymbol{R}}_2$ is the relative separation of their centroids*. Note the antisymmetrization is carried out within the $\alpha$-particles but not between them. $f(R)$ gives the dependence on $|\boldsymbol{R}|$ and is the same for all angular momentum values; this is what we mean by a rigid rotator. As a definite ansatz for calculations with (IV,63) let us take oscillator wave functions for the internal functions $\phi_0$, and take

$$f(R) = R^4 \mathrm{e}^{-aR^2}\,. \qquad (IV,63\,a)$$

If we compute energy expectation values with these rigid rotator functions, we find they have identical potential energies, while their kinetic energies increase with increasing $l$ so as to give the characteristic $l(l+1)\,\hbar^2/2I$ rotational spacing.

A correct explanation of these levels must take into account the antisymmetrization between the $\alpha$-particles, that is, the $\alpha$-particles must be treated as $a$-clusters [21]. Let us use for this purpose the oscillator cluster functions for $Be^8$ found previously (eq. (III,15)).

$$\psi_{\rm CL}(Be^8) = \mathscr{A}\{\phi_0(\underset{\sim}{a_1})\,\phi_0(\underset{\sim}{a_2})\,\chi_{4\hbar\omega,l,m}(\boldsymbol{R})\} \qquad (IV,64)$$

where

$$\chi_{4\hbar\omega,l,m} = (R^4 + bR^2 + c)\,Y_{lm}(\theta_R,\varphi_R)\,\mathrm{e}^{-aR^2}\,. \qquad (IV,64\,a)$$

In the cluster function (IV,64) the antisymmetrization $\mathscr{A}$ goes over all nucleon coordinates. Hence in the relative motion functions $\chi$ the $bR^2$ and constant terms are of no consequence; they vanish under $\mathscr{A}$ because they correspond to low relative oscillations of the $a$-clusters. The result is then that if the rigid rotator functions (IV,63 and 63a) were totally antisymmetrized, i.e. including antisymmetrization between the $\alpha$-particles, they would be identical to the corresponding oscillator cluster functions (IV,64 and 64a). Despite this apparent similarity, the totally antisymmetrized cluster functions lead to a completely different explanation of the energy differences between the levels. We have already discussed the properties of these cluster functions; we know that they all have the same kinetic energy but are split by differing potential energy,

---

* We assume the spin and isospin dependence included in the internal cluster functions in this discussion. Note for $a$-clusters the total spin and isospin are both zero.

$l = 0, 2, 4$ in order of increasing energy. Hence the failure to anti-symmetrize between the clusters in the rigid rotator functions shifts the energy splitting completely from the potential to the kinetic energy. Antisymmetrization has the effect of lessening the kinetic energy splitting and increasing the potential energy splitting[*].

For a more precise investigation of the behaviour of the energy spectrum caused by the transition from distinguishable to completely indistinguishable particles we rewrite the $Be^8$ rotational wave functions by introducing parameter coordinates analogously to, and with the notations of eq. (IV,20). We begin with the rigid rotator function (IV,63):

$$\psi_{\text{rot}}(Be^8) = \int (\mathscr{A}\,\phi_0(a_1')) \, (\mathscr{A}\,\phi_0(a_2')) \, f(R') \, Y_{lm}(\theta_R', \varphi_R') \times$$
$$\times \, [\delta(\underset{\sim}{r_1} - \underset{\sim}{r_1'}) \, \delta(\underset{\sim}{r_2} - \underset{\sim}{r_2'}) \ldots \delta(\underset{\sim}{r_8} - \underset{\sim}{r_8'})] \, d\tau' . \tag{IV,65}$$

Now we substitute (primed) cluster relative and center of mass co-ordinates for the primed parameter coordinates and carry out the inte-grations and summation over all parameter variables except the angular orientation of $R'$, and write the result as

$$\psi_{\text{rot}}(Be^8) = \int \psi_{\text{int}}(r_i, \theta_R', \varphi_R') \, Y_{lm}(\theta_R' \varphi_R') \, d\Omega'$$
$$d\Omega' = \sin\theta_R' \, d\theta_R' \, d\varphi_R' . \tag{IV,66}$$

The $Y_{lm}(\theta_R', \varphi_R')$ are normalized spherical harmonics of the parameter arguments $\theta_R', \varphi_R'$. $\psi_{\text{int}}(r_i, \theta_R', \varphi_R')$ describes the nucleon configuration around the symmetry axis given by $(\theta_R', \varphi_R')$ independent of the direction of the axis. Moreover $\psi_{\text{int}}(r_i, \theta_R', \varphi_R')$ is rotationally invariant around the symmetry axis.

We now substitute (unprimed) cluster coordinates $R, \theta_R, \varphi_R, \bar{r}_i$ for the unprimed nucleon variables $\underset{\sim}{r_i}$ in $\psi_{\text{int}}(r_i, \theta_R', \varphi_R')$.

It can be shown easily that in the case where one does not anti-symmetrize between the two $a$-clusters, $\psi_{\text{int}}(r_i, \theta_R', \varphi_R')$ reduces to the following form

where

$$\psi_{\text{int}}(\underset{\sim}{r_i}, \theta_R', \varphi_R') = \psi_{\text{int}}^{\text{rot}} \equiv g(\bar{r}_i, R) \, \frac{\delta(\theta_R - \theta_R') \, \delta(\varphi_R - \varphi_R')}{\sin\theta_R}$$

$$\tag{IV,67}$$

$$g(\bar{r}_i, R) = e^{-\frac{a}{2}\sum\limits_{i=1}^{4} \bar{r}_i^2} \, e^{-\frac{a}{2}\sum\limits_{i=1}^{8} \bar{r}_i^2} \, R^4 \, e^{-a R^2} \, \xi(s_1 \ldots t_8) .$$

The most important point to note about eq. (IV, 67) is the $\delta$-function dependence on $\theta_R'$ and $\varphi_R'$. This means for instance that $\psi_{\text{int}}(r_i, \theta_R', \varphi_R')$ and $\psi_{\text{int}}(r_i, \theta_R'', \varphi_R'')$ are orthogonal for $\theta_R' \neq \theta_R''$ or $\varphi_R' \neq \varphi_R''$ when inte-grated over the nucleon coordinates $\underset{\sim}{r_i}$. This $\delta$-function dependence and consequent orthogonality is the result of our failure to antisymmetrize between the $a$-clusters, so that eq. (IV,67) is valid only for the rigid rotator function.

---

[*] In the case of $Be^8$ we have seen this takes place in such a way that the ratio $\dfrac{E_{l=4} - E_{l=0}}{E_{l=2} - E_{l=0}} = \dfrac{10}{3}$ is preserved .

The expectation value for the energy has the following general form for Be$^8$ wave functions $\psi$ written in the parameter notation see (IV,9) and (IV,9a)

$$\langle E \rangle = \frac{\langle \psi \mid \overline{T} + V \mid \psi \rangle}{\langle \psi \mid \psi \rangle}$$

$$= \frac{\int\int Y^*_{lm}(\theta''_R, \varphi''_R) \, f(\gamma) \, Y_{lm}(\theta'_R, \varphi'_R) \, d\Omega'' \, d\Omega'}{\int\int Y^*_{lm}(\theta''_R, \varphi''_R) \, g(\gamma) \, Y_{lm}(\theta'_R, \varphi'_R) d\Omega'' \, d\Omega'} \qquad \text{(IV,68)}$$

with

$$f(\gamma) = \langle \psi_{\text{int}}(\underset{\sim}{r}_i, \theta''_R, \varphi''_R) \mid \overline{T} + V \mid \psi_{\text{int}}(\underset{\sim}{r}_i, \theta'_R, \varphi'_R) \rangle$$

$$g(\gamma) = \langle \psi_{\text{int}}(\underset{\sim}{r}_i, \theta''_R, \varphi''_R) \mid \psi_{\text{int}}(\underset{\sim}{r}_i, \theta'_R, \varphi'_R) \rangle \qquad \text{(IV,69)}$$

where $\gamma$ is the angle between the directions $(\theta''_R, \varphi''_R)$ and $(\theta'_R, \varphi'_R)$. We have already split off the center of mass dependence; hence we consider only the relative kinetic energy, $\overline{T}$. That the expression (IV,69) depends only on the angle $\gamma$ between the directions $(\theta''_R, \varphi''_R)$ and $(\theta'_R, \varphi'_R)$ comes from the fact that $\psi_{\text{int}}$ is rotationally invariant around the symmetry axis and the energy-operator is itself rotationally invariant.

To calculate the energy expectation values of the rigid rotator functions we substitute (IV,67) into (IV,69), whereby we obtain for $f(\gamma)$ and $g(\gamma)$:

$$f(\gamma) = \left\{ E_{\text{int}} + \frac{\hbar^2}{4M} \left\langle \frac{1}{R^2} \right\rangle \left[ \frac{1}{\sin\theta''_R} \frac{\partial}{\partial\theta''_R} \left( \sin\theta''_R \frac{\partial}{\partial\theta''_R} \right) + \frac{1}{\sin^2\theta''_R} \frac{\partial^2}{\partial\varphi''^2_R} \right] \right\} \times$$

$$\times \frac{\delta(\theta'_R - \theta''_R)\,(\varphi'_R - \varphi''_R)}{\sin\theta''_R} \qquad \text{(IV,70)}$$

$$= \sum_l \left\{ E_{\text{int}} + \frac{\hbar^2}{4M} \left\langle \frac{1}{R^2} \right\rangle l(l+1) \right\} \sum_{m=-l}^{l} Y^*_{lm}(\theta''_R \, \varphi''_R) \, Y_{lm}(\theta'_R \, \varphi'_R)$$

and

$$g(\gamma) = \frac{\delta(\theta''_R - \theta'_R)\,\delta(\varphi''_R - \varphi'_R)}{\sin\theta''_R} = \frac{1}{2\pi} \frac{\delta(\gamma)}{\sin\gamma}$$

$$\qquad \text{(IV,70a)}$$

$$= \sum_l \sum_{m=-l}^{l} Y^*_{lm}(\theta''_R \, \varphi''_R) \, Y_{lm}(\theta'_R \, \varphi'_R) \, .$$

$E_{\text{int}}$ is the internal kinetic and potential energy of Be$^8$ without its rotational energy. The final expressions for $f(\gamma)$ and $g(\gamma)$ in (IV,70) and (IV,70a) can be derived from the initial expressions most simply by expanding them, with the help of the addition theorem for spherical harmonics, in products of spherical harmonics, as the final expressions of (IV,70) and (IV,70a) indicate.

If one substitutes (IV,70) and (IV,70a) in (IV,68) then one obtains for the energies of the Be$^8$ levels

$$E_l = E_{\text{int}} + \frac{2\hbar^2}{AM} \frac{1}{\langle R^2 \rangle} l(l+1); \quad l = 0, 1, 2, \ldots \qquad \text{(IV,71)}$$

where $A = 8$ is the atomic number. $E_{\text{int}}$ is independent of $l$. The second term is the rotational energy which is of purely kinetic nature. In contradiction to experiment this rotational structure remains valid for all $l$ values; it does not cut off above $l = 4$, and includes unobserved odd $l$

values. Mathematically these results follow from the fact that $f(\gamma)$ and $g(\gamma)$ are superpositions of $\delta$-functions and their derivatives, which in turn follows from the $\delta$-function nature of $\psi_{\text{int}}$ as a function of $\theta'_R$ and $\varphi'_R$, seen in (IV,67).

Let us now perform the calculation for $\psi_{\text{CL}}(Be^8)$, where the antisymmetrization is performed over all nucleon coordinates, and compare it step by step with the rigid rotator calculation. We begin the same way by introducing cluster parameter coordinates and integrating over all parameter coordinates except $\theta'_R$ and $\varphi'_R$ (see for instance (IV,24) and (IV,26)):

$$\psi_{\text{CL}}(Be^8) = \int \{\phi_0(\underset{\sim}{a'_1})\,\phi_0(\underset{\sim}{a'_2})\,\chi(\boldsymbol{R'})\}\,\mathscr{A}\,[\delta(\underset{\sim}{r_1} - \underset{\sim}{r'_1})\ldots\delta(\underset{\sim}{r_8} - \underset{\sim}{r'_8})]\,d\tau'$$

$$= \int \psi_{\text{int}}(\underset{\sim}{r_i},\theta'_R,\varphi'_R)\,Y_{lm}(\theta'_R,\varphi'_R)\,d\Omega' \qquad \text{(IV,72)}$$

where now all nucleon coordinates in $\psi_{i\ t}(\underset{\sim}{r_i},\theta'_R,\varphi'_R)$ are indistinguishable. Again $\psi_{\text{int}}$ has rotational symmetry about the $(\theta'_R,\varphi'_R)$ axis. Moreover, when we antisymmetrize between clusters, $\psi_{\text{int}}$ must have positive parity, so that odd $l$ states vanish automatically in the above integral*.

When we introduce cluster coordinates for the (unprimed) nucleon coordinates all parallelism between the two calculations ends. For the exchange terms between the two $\alpha$-clusters destroy the $\delta$-function relation between $\theta_R$ and $\theta'_R$ and $\varphi_R$ and $\varphi'_R$. This reflects the fact that with completely indistinguishable nucleons we can no longer know which nucleons belong to which cluster; hence the centers of mass of the clusters are indeterminate, which is why we introduce the parameter coordinates in the first place. The functional dependence of $\psi_{\text{int}}(\underset{\sim}{r_i},\theta'_R,\varphi'_R)$ on $\theta'_R$ and $\varphi'_R$ is now much more complicated, and in particular $\psi_{\text{int}}(\underset{\sim}{r_i},\theta'_R,\varphi'_R)$ and $\psi_{\text{int}}(\underset{\sim}{r_i},\theta''_R,\varphi''_R)$ are no longer orthogonal if $(\theta'_R,\varphi'_R)$ and $(\theta''_R,\varphi''_R)$ do not coincide.

To calculate the energy expectation value of the oscillator cluster functions (IV,72) we again employ (IV,68) and (IV,69). These equations remain valid because of the symmetry character of $\psi_{\text{int}}$. To continue the calculation one has to introduce specific two-body interaction forces, such as (IV,14). If one assumes for $V_{ik}$ a Gaussian form then the integrals can be performed explicitly. Since the details are quite lengthy and have already been reported elsewhere, [21], we shall just summarize the salient results here which have general application to even-even rotational states and which show the influence of complete antisymmetrization.

It turns out for the three lowest states of $Be^8$ described by (IV,72) the kinetic energy is the same and the potential energy is proportional to $l(l+1)$. This agrees with our earlier general remarks in the frame of the oscillator cluster model (chap. III sect. C). The result for the poten-

---

* One can say in this behaviour of the $\alpha$-particle states of $Be^8$ the boson character of the $\alpha$-particles is expressed. Note, by positive parity of $\psi_{\text{int}}(\underset{\sim}{r_i}\theta'_R\varphi'_R)$ we mean $\psi_{\text{int}}(-\underset{\sim}{r_i},\theta'_R,\varphi'_R) = \psi_{\text{int}}(\underset{\sim}{r_i},\theta'_R,\varphi'_R)$. Alternatively this may be expressed $\psi_{\text{int}}(\underset{\sim}{r_i},\pi - \theta'_R,\pi + \varphi'_R) = \psi_{\text{int}}(\underset{\sim}{r_i},\theta'_R,\varphi'_R)$.

tial energy is independent of the assumed two-nucleon forces, [22], [23], because any such kind of forces can be approximated by a superposition of forces of Gaussian form. It can be shown further that for $l > 4$ the rotational structure of the Be[8] spectrum breaks down, because the kinetic energy then becomes a function of $l$ and the $l(l + 1)$ structure of the potential energy is altered. Hence the rotational structure of Be[8] is quite accidental.

While the detailed results above apply only for the special wave functions with which we describe Be[8] here, there is a very general reason for the shift of the energy splitting from the kinetic to the potential energy. Antisymmetrization between clusters destroys the possibility of defining the symmetry axis in terms of the nucleon coordinates. Hence $\psi_{\text{int}}(\mathbf{r}_i, \theta'_R, \varphi'_R)$ is no longer orthogonal for different directions of $(\theta'_R, \varphi'_R)$. Hence $f(\gamma)$ and $g(\gamma)$ no longer vanish for $\gamma \neq 0$, as is the case when we do not antisymmetrize between the clusters, with the further consequence that the kinetic energy loses its rotational structure and the potential energy becomes $l$ dependent. This energy shift due to complete antisymmetrization is quite general and not restricted to our Be[8] example, as we will discuss immediately.

A generalization of formula (IV,72) will now serve as a starting point for a more general investigation of nuclear rotational states. For the moment we restrict ourselves to states of even-even nuclei in which the nucleon spins cancel to zero; this is true at least in the ground state of all stable even-even nuclei. If the cluster structure of such a nucleus is given, it is always possible to group the clusters into larger clusters until finally only two remain. These two clusters may be different from each other. In the most general case one can have a linear superposition of such two cluster configurations. If all these two cluster configurations are given a common relative angular momentum and if we again use our parameter representation, we can describe the wave function $\psi$ of the nucleus as a superposition of internal wave functions, i.e. analogously to (IV,72)

$$\psi = \int \psi_{\text{int}}(\mathbf{r}_i, \theta', \varphi') \, Y_{lm}(\theta', \varphi') \, \mathrm{d}\Omega' . \tag{IV,73}$$

The parameter coordinates $(\theta', \varphi')$ define the direction of separation of the centroids of the two cluster configurations.

We introduce a further restriction. We demand that $\psi_{\text{int}}(\mathbf{r}_i, \theta', \varphi')$ be rotationally invariant about the $(\theta', \varphi')$ axis. This can be achieved in several ways. One way for even-even nuclei is to take all clusters in the ultimate two cluster superposition spherically symmetric about their respective centers if mass, that is, the clusters may possibly be internally excited but their internal angular momentum must be zero. For definiteness we shall only consider such states of even-even nuclei for which $(\theta', \varphi')$ can be made a symmetry axis in this way. These restrictions can be somewhat relaxed after the general direction of the calculation is made clear. For the even-even nuclear states now considered $l$ and $m$ in (IV,73) refer to the total angular momentum of the nucleons and will be zero in the ground state. Because even-even nuclei have positive parity in the ground state, $\psi_{int}$ must have positive parity in the rotational states built on the ground state.

With the wave functions (IV,73) it now follows from our symmetry conditions, analogously to the Be$^8$ case, that the expectation value of the energy must again have the form of (IV,68)

$$\langle E \rangle = \frac{\int Y^*_{lm}(\theta'', \varphi'')\, f(\gamma)\, Y_{lm}(\theta', \varphi')\, d\Omega''\, d\Omega'}{\int Y^*_{lm}(\theta'', \varphi'')\, g(\gamma)\, Y_{lm}(\theta', \varphi')\, d\Omega''\, d\Omega'} \qquad \text{(IV,68a)}$$

where $f(\gamma)$ and $g(\gamma)$ are defined in the same way as in (IV,69). We now define rotational states by the requirement that $\psi_{int}(\mathbf{r}_i, \theta', \varphi')$ in (IV,73), and therefore $f(\gamma)$ and $g(\gamma)$ are independent of $l$ and $\widetilde{m}$. Further due to the positive parity of $\psi_{int}(\mathbf{r}_i, \theta', \varphi')$ only even $l$ values can appear.

If one does not antisymmetrize the ultimate two cluster configurations, then $f(\gamma)$ and $g(\gamma)$ assume the same form as in the rigid rotator Be$^8$ case (see eqs. (IV,70) and (IV,70a)), that is: they are composed of $\delta$-functions and derivatives of $\delta$-functions. Therefore we obtain an energy spectrum of rotational structure for the rotational states, and the rotational excitation energy is purely kinetic.

If we completely antisymmetrize, $f(\gamma)$ and $g(\gamma)$ will again become different from zero for $\gamma \neq 0$. Therefore rotational states as defined above will in general not have an energy spectrum of pure rotational structure. However, under proper circumstances $f(\gamma)$ and $g(\gamma)$ will be sharply peaked about $\gamma = 0$ and will be small elsewhere[*].

If the nucleus is elongated, i.e. if $\psi_{int}(\mathbf{r}_i, \theta', \varphi')$ is elongated along the $(\theta', \varphi')$ axis, then the overlap between $\psi_{int}(\mathbf{r}_i, \theta'', \varphi'')$ and $\psi_{int}(\mathbf{r}_i, \theta', \varphi)$ will be reduced when $(\theta', \varphi')$ and $(\theta'', \varphi'')$ are not alligned. In fact, roughly speaking any slight reduction in the overlap for a single nucleon will be raised to the power of the number of nucleons which contribute to the deformation and hence will lead to a greatly reduced value for the complete overlap integral. We can therefore see that the peaking of $f(\gamma)$ and $g(\gamma)$ is most likely to be pronounced when $A$ is large and when the nucleus has its last shell roughly half filled so as to produce a large deformation, [49, 50].

We now use these properties of $f(\gamma)$ and $g(\gamma)$ for highly deformed heavy even-even nuclei for a more detailed investigation of their low energy spectra. For this we expand again $f(\gamma)$ and $g(\gamma)$ in spherical harmonics. At first we obtain

$$f(\gamma) = \sum_l f_l Y_{l,0}(\gamma)$$

and                                                                                      (IV,74)

$$g(\gamma) = \sum_l g_l Y_{l,0}(\gamma)$$

---

[*] From the parity of $\psi_{int}$ it follows that $f(\gamma)$ and $g(\gamma)$ must also be sharply peaked at $\gamma = \pi$, but this does not affect our general results.

with

$$f_l = 2\pi \int_0^\pi f(\gamma) \, Y_{l0}(\gamma) \, \sin\gamma \, d\gamma$$

$$g_l = 2\pi \int_0^\pi g(\gamma) \, Y_{l0}(\gamma) \, \sin\gamma \, d\gamma \,. \tag{IV,74a}$$

By use of the addition theorem for spherical harmonics which states that

$$Y_{l,0}(\gamma) = 2\sqrt{\frac{\pi}{2l+1}} \sum_{m=-l}^{l} Y_{l,m}^*(\theta'', \varphi'') \, Y_{l,m}(\theta', \varphi) \tag{IV,74b}$$

we obtain for the expectation value of the energy, if we introduce (IV,74) in (IV,73)

$$\langle E \rangle_{lm} = \frac{f_l}{g_l} \,. \tag{IV,75}$$

So far the calculation is exact.

For the calculation of $f_l$ and $g_l$ we expand $Y_{l,0}(\gamma) \sin\gamma$ in a power series in $\gamma$

$$Y_{l,0}(\gamma) \sin\gamma = \frac{1}{2}\sqrt{\frac{2l+1}{\pi}} \left( \gamma - l(l+1)\,\gamma^3 - \frac{\gamma^3}{6} + \text{higher terms} \right).$$

Substituting this in (IV,74a) using the sharp peaking of $f(\gamma)$ and $g(\gamma)$, we need retain only the first few terms in the integrations. Substituting the result into (IV,75) we obtain

$$\langle E \rangle_{lm} = A + B\,l(l+1) \tag{IV,76}$$

$A$ and $B$ depend on the special ansatz for the two-body forces and on the structure of $\psi_{int}$, but are independent of $l$.

Thus we obtain a set of energy levels with the characteristic rotational spacing as a function of angular momentum. But whereas in the simple rotator model (no antisymmetrization between the two ultimate clusters) the $l(l+1)$ term arises entirely from kinetic energy, in the present anti-symmetrized cluster wave functions it has in general a contribution from the potential energy also. This gives us insight why the old attempts to calculate the moment of inertia of a nucleus to fit the rotational levels were unsuccessful. It can be shown from the present treatment that the value of $\hbar^2/2B$ above, corresponding to the moment of inertia, must lie between the moment of inertia calculated from the rigid rotator model and the irrotational flow model (see for this the remarks in [21] p. 464). Precise numerical calculations of the moment of inertia are complicated due to the antisymmetrization; however see R. PEIERLS and I. YOCCOZ [54], I. YOCCOZ [55].

We must now inquire whether the rotational wave functions (IV,72) are good representations of the low energy states of deformed even-even nuclei. Our approach will again be based on the Ritz variational principle. In the range of high $A$ and with the last shell partially filled, the remarks above indicate that the spacing between the energy expectation values of successive wave functions (IV,72) should be of the same order of magnitude as that resulting from the conventional kinematic rotator theory ($\sim$ 100 MeV). On the other hand, due to the short range of the nucleon-nucleon interaction, a change in the internal structure of the nucleus or a surface change will greatly alter the energy expectation value, usually by at least one MeV. In general, under these conditions the rotational wave functions (IV,72) will represent in good approximation the true nuclear eigenfunctions for heavy deformed nuclei of low energies ($<$ 1 MeV); admixture of other amplitudes corresponding to internal oscillation excitations of the nuclei, and therefore high energy expectation values, will in general be small. A rich rotational spectrum is in fact observed experimentally in these nuclei. Near closed shells the rotational structure gradually vanishes in agreement with our arguments.

At low $A$ our argument that $f(\gamma)$ and $g(\gamma)$ are sharply peaked at $\gamma = 0$ breaks down and we can no longer make the general conclusion that the energy expectation values of the rotational (as we defined it by its structure) wave function (IV,73) have a rather exact $l(l + 1)$ form with narrow spacing. For example, the rotational wave functions (IV,73) give a rather good representation of some of the relatively low excited states of Ne[20]; that is, if $\psi_{\text{int}}(\underset{\sim}{r_i}, \theta', \varphi')$ is independent of $l$ and $m$, and we take successive even values of $l$ in the spherical harmonic $Y_{lm}(\theta', \varphi')$, then (IV,73) describes certain low levels of Ne[20]. By our definition these are rotational levels. However their energy spacing has only in a very rough approximation the $l(l + 1)$ form.

With the rotational wave functions (IV,73) it is also possible, for the case of heavy deformed even-even nuclei, to predict correctly the ratios of the $\beta$ and $\gamma$ transition probabilities for states of different angular momentum in a given rotational series. For instance one obtains for electric quadrupole transitions the following ratio, [21],

$$\left| \frac{M_l(l + 2, l + 2 \to l, l)}{M_l(l, l \to l - 2, l - 2)} \right|^2 = \frac{(l + 1)\,(l + 2)\,(2l - 1)\,(2l + 1)}{(2l + 3)\,(2l + 5)\,(l - 1)} . \quad \text{(IV,77)}$$

This is exactly the angular momentum dependence which one expects from the phenomenological collective model, [49, 50]. However one obtains the absolute values of the transition probabilities given by the collective model only if the nucleons are treated as distinguishable, in other words, if the exchange terms are neglected and only the direct terms remain. As can be shown, the indistinguishability of the nucleons reduces these absolute values but in such a way that the ratio (IV,77) is unaffected. Due to the antisymmetrization the explicit calculation of the absolute values of such transition probabilities again becomes complicated.

In a similar way as was sketched for the even-even nuclei with ground state spin 0, one can investigate the structure of negative parity rotational bands of even-even nuclei and their $\beta$ and $\gamma$ transition probabilities. Here the internal wave functions have negative parity and therefore only odd angular momentum values will appear in the rotational states*, see for instance K. ALDER et al. [56].

The presence of clusters with internal spin does not greatly affect our definition of rotational states except that now, due to the spin-orbit forces, the total spin of the clusters can couple to the orbital angular momentum to split a rotational level of a given $l$ value. Thus for instance one may consider the low excited $T = 0$ states ($I = 3^+$, $2^+$, $1^+$; $a - d$ cluster structure) of $Li^6$ as a triplet (*deuteron* cluster has spin one) splitting of the $l = 2$ rotational level built on the ground state ($I = 1^+$, $l = 0$). Similarly the low excited positive parity states $\left(I = \dfrac{5^+}{2}, \dfrac{3^+}{2},\right.$ $\dfrac{9^+}{2}, \dfrac{7^+}{2}$; $O^{16} - t$ cluster structure$\Big)$ of $F^{19}$ are the doublet split $\Big($*triton* cluster has spin $\dfrac{1}{2}\Big)$ rotational levels $l = 2$, $l = 4$, built on the ground state $\left(I = \dfrac{1^+}{2}, l = 0\right)$. In the same way a set of doublet split negative parity rotational levels of even $l$ can be built on the first excited state $\left(I = \dfrac{1^-}{2}, l = 0; N^{15} - \alpha \text{ cluster structure}\right)$ of $F^{19}$. The negative parity arises from the internal parity of the $N^{15}$-cluster; however as this is an odd-even nucleus the integral (IV,73) does not vanish for even $l$ values as in the case of even-even nuclei negative parity levels (see below). We want to point out that in the frame of the oscillator assumptions, after anti-symmetrizations, the states described here are essentially equal to the corresponding supermultiplet states, WIGNER [57], the shell model states with interaction, D. INGLIS [58], I. ELLIOT and B. FLOWERS [59], and the $SU_3$-states, I. ELLIOT [60], A. BOHR and B. BAYMAN [60].

The parameter method used above can also be applied to the investigation of rotational levels of heavy odd-even nuclei. One main difference to even-even nuclei comes from the fact that with an odd number of nucleons $\psi_{\text{int}}(r_i, \theta', \varphi')$ is no longer rotationally invariant around the direction $(\theta', \widetilde{\varphi'})$, with the further consequence that now odd and even $l$ values can appear in the rotational spectra of odd-even nuclei. More generally the parameter method can be applied to investigate any sort of collective motion, with the appropriate collective coordinates being explicitly set off in parameter coordinates; we could for instance handle oscillational collective states, [50], in this way.

It should be mentioned that our definition of rotational states does not lead to a unique specification of a given excited rotational state. For example if $f(r_i, \theta', \varphi')$ has the property

$$\int f(r_i, \theta', \varphi')\, Y_{lm}(\theta', \varphi')\, d\Omega' = 0 \qquad\qquad \text{(IV,78)}$$

---

* The internal angular momenta of the cluster wave functions which form the internal wave functions must again be zero. The behaviour of such negative parity internal wave functions can be studied in detail in the low excited negative parity states of $O^{16}$, see chap. III sec. E.

for some set of $l$ values, say $l = 0, 2, 4$, then the wave function

$$\bar{\psi} = \int \{\psi_{\text{int}}(\underset{\sim}{r_i}, \theta', \varphi') + f(\underset{\sim}{r_i}, \theta', \varphi')\} Y_{lm}(\theta', \varphi') \, d\Omega$$

will give the same rotational states as (IV,73) for the $l = 0$ ground state, the $l = 2$ states, and so on, but will give different states for $l$ values such that

$$\int f(\underset{\sim}{r_i}, \theta', \varphi') Y_{lm}(\theta', \varphi') \, d\Omega' \neq 0 . \qquad (\text{IV,78a})$$

The extra degrees of freedom of this type can be removed at least in principle by applying the variational criterion for the energies of higher rotational states.

Finally we want to make some general remarks about the connection between our treatment of rotational states and the somewhat similar method of Yoccuz and Peierls, [54, 55]. They construct their internal wave functions from a rotationally symmetric potential in which the nucleons are moving. From this it follows that the real symmetry axis of their internal wave functions oscillates around the symmetry axis of their given potential. Therefore in practical cases it is very difficult to split off the center of mass motion of the whole nucleus in a clear way. Only if the deformation of the potential and the number of nucleons in the nucleus is large will this oscillation be negligible to a first approximation. Therefore this method is not very suitable for constructing rotational wave functions for nuclei with relatively small deformations or for light nuclei. In the method discussed above we do not have this trouble. The symmetry axis of the internal wave function was fixed directly and therefore the center of mass motion can be explicitly split off. The whole uncertainty of the direction of the nuclear symmetry axis is described by the spherical harmonics $Y_{lm}(\theta', \varphi')$. The oscillation of the internal symmetry axis in the method of Yoccuz and Peierls appears also in the fact that in this method, in contrast to ours, the internal wave functions $\psi_{\text{int}}(\underset{\sim}{r_i}, \theta', \varphi')$ overlap for values of $(\theta', \varphi')$ specifying different directions of the rotationally symmetric potential even when the nucleons are considered as distinguishable. It can be shown that for large deformations and for many nucleons in the nuclei the results of both methods approximate each other. For instance the deviation in the rotational energies are of the order of magnitude $(\Delta R/R)^2 \langle E \rangle_{lm}$, as a rough estimation shows; $\Delta R$ is the amplitude of the zero point oscillation of the center of mass of the whole nucleus in the given rotationally symmetric potential and $R$ is the nuclear radius. $(\Delta R/R)^2$ is roughly proportional to $1/A$. This mutual approach of both methods makes the method of Yoccuz and Peierls very useful for calculating the moments of inertia of heavy nuclei approximately, because it is much easier in the case of many nucleon coordinates to carry out integrals over single particle wave functions than over cluster wave functions with their different relative and center of mass cluster coordinates.

# V. Nuclear Reactions

## A. General considerations

At the beginning of the last chapter was discussed how the oscillator cluster wave functions may be generalized to arrive at better variational wave functions for the description of nuclear bound states. The main feature of the generalization was to treat the internal frequencies in the clusters and the frequencies of the relative motion of the clusters as independent variational parameters. We mentioned then that this was the first step to connect nuclear structure theory with nuclear reaction theory in a very natural way.

To obtain the connection we make a further generalization of our cluster functions, in particular of the relative motion part of the cluster function. We begin with a given generalized cluster function. To make the new generalization we hold the given internal functions of the clusters fixed, but we now vary the relative motion function completely arbitrarily. The relative motion function $\chi(\mathbf{R})$ may thus be written $\chi(\mathbf{R}) + \delta\chi(\mathbf{R})$ upon variation, where $\delta\chi$ is an arbitrary function; alternatively we may say that $\chi(\mathbf{R})$ is now given an infinite number of variational parameters. We then demand that the expectation value of the nuclear Hamiltonian be stationary* under this variation, $\delta\langle H\rangle = 0$, and from this condition derive the equations** determining $\chi(\mathbf{R})$. That we only demand that $\langle H\rangle$ be stationary and not necessarily a minimum means that we can handle both bound states and scattering and reaction states, depending on the asymptotic boundary conditions we impose on $\chi(\mathbf{R})$. In the most general description we have a superposition of cluster functions, with all relative motion functions varied, so that any bound state configuration or any number of reaction channels may be considered. If we demand as asymptotic conditions for the relative motion functions that they vanish exponentially for infinite separation distances of the clusters, then the set of equations which we obtain from the variational principle can be used for treating bound states of the nuclei. On the other hand, if we demand that the relative functions behave asymptotically like a superposition of incoming and outgoing waves, then we can use this set of equations for the description of nuclear scattering and reaction processes. The connection between these bound state wave functions and scattering or reaction process wave functions is completely analogous to the connection between the bound state wave functions of the hydrogen atom and the electron-proton Rutherford scattering wave functions. To make our general remarks clear and to come to a quantitative formulation we consider some specific examples.

---

* The constraint of normalization of the wave function is again taken into account by a Lagrange multiplier as will be discussed immediately.

** To derive equations for the approximate solution of the many particle Schrödinger equation from a variational principle using a limited ansatz for the trial wave function is a standard mathematical procedure, c.f. Hartree-Fock equations.

## B. Derivation of coupled channel equations for nuclear reactions and nuclear bound states

At first we will discuss a two cluster problem, for instance the bound states and the scattering states of an $\alpha$-cluster and *triton*-cluster or an $\alpha$-cluster and an $O^{16}$-cluster and so on. For simplicity we neglect for the moment any internal excitation or distortion of the clusters and all spin-orbit coupling effects*. We begin with the usual ansatz for the wave function of our system.

$$\psi = \mathscr{A}\{\phi_0(1)\ \phi_0(2)\ \chi(\boldsymbol{R})\} \qquad (V,1)$$

$\phi_0(1)$ and $\phi_0(2)$ are the internal functions for the two clusters including spin and isospin dependence, and in the present generalization are taken as given functions not subject to variation. However, the relative motion function $\chi(\boldsymbol{R})$ in the present generalization is taken as a completely free variational wave function and will not, as was done until now, be approximated by a given function containing one or a restricted number of variational parameters. For a precise definition of what function we have to vary in our variational procedure it is again useful to introduce parameter coordinates. Therefore we write

$$\psi = \int \phi_0(1')\ \phi_0(2')\ \chi(\boldsymbol{R'})\ \mathscr{A}\left[\delta(\boldsymbol{r_1} - \boldsymbol{r_1'}) \cdots \delta(\boldsymbol{r_A} - \boldsymbol{r_A'})\right] \mathrm{d}\tau' \qquad (V,2)$$

where $\chi(\boldsymbol{R'})$ is the variational function which has to be arbitrarily varied at every point $\boldsymbol{R'} = \boldsymbol{R_1'} - \boldsymbol{R_2'}$. As pointed out in chapter II we may now perform a straightforward variational process to obtain the equation determining our variational function $\chi(\boldsymbol{R'})$. Thus setting the variation of the energy expectation value to zero, we have

$$0 = \delta\langle \bar{H} \rangle = \delta\langle \psi | \bar{H} + \lambda | \psi \rangle$$
$$= A!\ \delta \int\!\!\int \phi_0^*(1'')\ \phi_0^*(2'')\ \chi^*(\boldsymbol{R''})\ [\mathscr{A}_{r'}\ \delta(\boldsymbol{r_1'} - \boldsymbol{r_1''}) \cdots \delta(\boldsymbol{r_A'} - \boldsymbol{r_A''})] \times$$
$$\times\ (\bar{H}(\boldsymbol{r_i'}\ \boldsymbol{p_i'}) + \lambda)\ \phi_0(1')\ \phi_0(2')\ \chi(\boldsymbol{R'})\ \mathrm{d}\tau'\ \mathrm{d}\tau'' \qquad (V,3)$$

In (V,3) we have already performed the sums and integrals over the (unprimed) nucleon coordinates. Because we split off the center of mass dependence in (V,1)** we neglect integration over the center of mass nucleon (unprimed) coordinate and use the intrinsic nuclear Hamiltonian in the center of mass system, see (IV,9) and (IV,9a),

$$\bar{H} = \bar{T} + V = \bar{T} + \sum_i^A \sum_{k\neq i}^A \frac{1}{2} V_{ik} \qquad (V,4)$$

Note that $\int \mathrm{d}\tau'$ and $\int \mathrm{d}\tau''$ still include integration over the respective center of mass parameter (primed) coordinates, $\boldsymbol{R_{CM}'}$ and $\boldsymbol{R_{CM}''}$. $\mathscr{A}_{r'}$ refers to the antisymmetrization of just the singly primed $\boldsymbol{r_i'}$ coordinates

---

* Exchange polarization effects are automatically included by the antisymmetrization of the wave function as discussed in chapter IV section D.

** The scrupulous reader may reinsert the center of mass function $Z(\boldsymbol{R_{CM}})$ in (V,1) and show that the center of mass energy can be subtracted from the Lagrange multiplier as an irrelevant constant in the variation (V,3).

in the bracketed $\delta$-functions. $\lambda$ is the Lagrange multiplier which is introduced to take into account the constraint of normalization of the wave function. In (V,3) we have already used that one need only perform the antisymmetrization either in $\psi^*$ or in $\psi$.

From our ansatz (V,2) the variation $\delta$ in (V,3) operates only on $\chi$ and $\chi^*$. Now $\delta\chi$ and $\delta\chi^*$ are arbitrary linearly independent variations, hence their coefficients must be separately set to zero. Equating the coefficient of $\delta\chi^*(\boldsymbol{R}')$ to zero, we have

$$\int\int \phi_0^*(1'') \,\phi_0^*(2'') \left[\mathscr{A}_{r'}\delta(\underset{\sim}{\boldsymbol{r}}_1' - \underset{\sim}{\boldsymbol{r}}_1'') \cdot \delta(\underset{\sim}{\boldsymbol{r}}_A' - \underset{\sim}{\boldsymbol{r}}_A'')\right] [\bar{H}(\underset{\sim}{\boldsymbol{r}}_i' \,\boldsymbol{p}_i') + \lambda] \times$$
$$\times \,\phi_0(1') \,\phi_0(2') \,\chi(\boldsymbol{R}') \,\mathrm{d}\tau' \,\mathrm{d}\bar\tau'' = 0 \,. \tag{V,5}$$

$\mathrm{d}\bar\tau''$ denotes that after introducing cluster and center of mass parameter coordinates we integrate and sum over all double-primed coordinates except the coordinate $\boldsymbol{R}'' = \boldsymbol{R}_1'' - \boldsymbol{R}_2''$. The integrations and summations over all single-primed and double-primed internal cluster coordinates can be carried out at least in principle because we have assumed that the internal cluster functions are given functions. The performance of the integrations and summations leads to an integro-differential equation for $\chi(\boldsymbol{R}')$ of the following general form:

$$\left\{-\frac{\hbar^2}{2\,\mu}\,\boldsymbol{V}_{\boldsymbol{R}'}^2 + V(\boldsymbol{R}')\right\} \chi(\boldsymbol{R}') + \int K(\boldsymbol{R}',\boldsymbol{R}'')\,\chi(\boldsymbol{R}'')\,\mathrm{d}\boldsymbol{R}'' = E\,\chi(\boldsymbol{R}') \tag{V,6}$$

which is the end result of our variational process. The expression in the bracket, with $V(\boldsymbol{R}')$ a real function of $\boldsymbol{R}'$, comes essentially from those terms of our wave function with no antisymmetrization between the clusters. The integral kernel $K(\boldsymbol{R}',\boldsymbol{R}'')$ is hermitian in $\boldsymbol{R}'$ and $\boldsymbol{R}''$, i.e. $K(\boldsymbol{R}',\boldsymbol{R}'') = K^*(\boldsymbol{R}'',\boldsymbol{R}')$ and contains the exchange effects between the clusters which come from the antisymmetrization and the exchange character of the nuclear forces. The energy

$$E = E_{\mathrm{TOT}} - (E_{\mathrm{bind}}(1) + E_{\mathrm{bind}}(2)) \tag{V,6a}$$

is in the case of a bound system the negative separation energy which one needs to split off the two clusters into two free particles. In the case of a scattering process $E$ represents the scattering relative kinetic energy of the two particles in the center of mass system. $E_{\mathrm{TOT}}$ in (V,6a) is equal to the negative value of the Lagrange multiplier $\lambda$ and excludes the energy of the center of mass $\boldsymbol{R}_{\mathrm{CM}}$ in the lab system. $\mu$ is the reduced mass of the two clusters, $M_1 M_2/(M_1 + M_2)$. We want to emphasize that $K(\boldsymbol{R}',\boldsymbol{R}'')$ is also a function of $E_{\mathrm{TOT}}$.

The reality of $V(\boldsymbol{R}')$ and the hermiticity of $K(\boldsymbol{R}',\boldsymbol{R}'')$ guarantees the current conservation law, i. e. that the current of the incoming scattering particles is the same as the current of the outgoing particles. Further, these hermitian properties guarantee that all solutions of (V,6) can be made orthogonal, which means that our approximate solutions to the many-body Schrödinger equation are also orthogonal.

We see from (V,6) that the bound states and scattering states of the considered two cluster system are described by a normal two-particle Schrödinger equation — in the relative motion coordinates of the clusters

— only that to the local interaction term of the clusters a nonlocal interaction term described by the integral kernel $K(\mathbf{R}', \mathbf{R}'')$ is added in which
all antisymmetrization effects and exchange effects are contained.

Due to the rotational invariance of the Hamiltonian (V,4) the poten-
tial $V(\mathbf{R}')$ can depend only on $R' = |\mathbf{R}'|$ and, analogously to our con-
siderations on the rotational states of even-even nuclei, $K(\mathbf{R}', \mathbf{R}'')$ can
depend only on $|\mathbf{R}'|$, $|\mathbf{R}''|$ and $\mathbf{R}' \cdot \mathbf{R}'' = R' R'' \cos\gamma$. Here $\gamma$ is again the
angle between the vector $\mathbf{R}'$ and the vector $\mathbf{R}''$. Therefore it is possible
analogously to (IV,74, 74a, 74b) to expand $K(\mathbf{R}', \mathbf{R}'')$ in a sum of products
of spherical harmonics.

$$K(\mathbf{R}', \mathbf{R}'') = \frac{1}{R'R''} \sum_l K_l(R', R'') \, Y_{l0}(\gamma) \tag{V,7}$$

$$= \frac{1}{R'R''} \sum_l \sum_{m=-l}^{l} K_l(R', R'') \sqrt{\frac{4\pi}{2l+1}} \, Y_{lm}^*(\theta'', \varphi'') \, Y_{lm}(\theta', \varphi')$$

with*

$$K_l(R', R'') = 2\pi R' R'' \int_0^\pi K(\mathbf{R}', \mathbf{R}'') \, Y_{l0}(\gamma) \sin\gamma \, d\gamma . \tag{V,7a}$$

If one expands $\chi(\mathbf{R}')$

$$\chi(\mathbf{R}') = \sum_{l,m} \frac{u_{lm}(R')}{R'} \, Y_{lm}(\theta', \varphi') \tag{V,7b}$$

then one obtains the following equation for $u_{lm}(R') = u_l(R')$

$$\left\{ -\frac{\hbar^2}{2\mu} \frac{d^2}{dR'^2} + \frac{\hbar^2 l(l+1)}{2\mu R'^2} + V(R') \right\} u_l(R') + \int_0^\infty K_l(R', R'') \, u_l(R'') \, dR'' \tag{V,8}$$

$$= E \, u_l(R') .$$

For the solution of (V,8) one has besides the asymptotic conditions
discussed before, the boundary condition $u_l(R' = 0) = 0$. One sees from
(V,8) that the integrodifferential equation (V,6) can be split off in a set of
integro-differential equations where every equation belongs to a definite
orbital angular momentum value $l$. This expresses nothing but the
conservation law of orbital angular momentum. Note we neglect spin-
orbit forces in $\overline{H}$. If one also takes into account the internal spin of the
clusters, then we can show completely analogously to the above that one
obtains for every total angular momentum value a separate integro-
differential equation**. The formulas become a little more complicated
however.

Next we consider a reaction process in which two different cluster
structures participate, for instance a process where the compound nucleus
can decay in two different reaction channels. An example of such a
process is the bombarding of $\alpha$-particles with neutrons of more than
17.6 MeV scattering energy in the center of mass system. This is the

---

* Examples of explicitly evaluated kernels $K_l(R', R'')$ for $\alpha - \alpha$ scattering can
be found in E. VAN DER SPUY [61].
** These equations can be split further, due to the reflection invariance of the
Hamiltonian, into equations with definite parity.

threshold energy for the decay of the $He^5$ compound nucleus into a triton and deuteron. Just above this energy at $17.8\,MeV$ center of mass scattering energy (or $16.7\,MeV$ above the $\frac{3^-}{2}$ $He^5$ resonance state) exists a $\frac{3^+}{2}$ resonance state which can be described in good approximation by a triton-deuteron cluster configuration, as already discussed in chapter IV section A. Therefore in the scattering energy region above about 17 MeV we must also take into account the triton-deuteron cluster structure and the triton-deuteron decay of the $He^5$ compound nucleus. Another example of this kind is the scattering of deuterons on $\alpha$-particles with more than $14.3\,MeV$ scattering energy in the center of mass system. This is the deuteron $\alpha$-particle threshold energy for the decay of the $Li^6$ compound nucleus into a triton and $He^3$.

Certainly in principle one could handle processes in which two different cluster structures participate with the complete set generated by a single cluster representation. But as mentioned in chapter IV section B, it is much more convenient to employ a mixed representation, because in a single cluster representation a second cluster structure is usually represented by an extremely complicated superposition of highly excited states of this first cluster representation, especially if the clusters of the second cluster structure are far apart. Usually one has to include continuum internal states of the clusters in the first cluster representation to represent scattering states of the second cluster structure.

Therefore to formulate such processes quantitatively we make the following "ansatz" for the variational wave function of the total system.

$$\psi = \mathscr{A} \left\{ \phi(1)\ \phi(2)\ \chi(\mathbf{R}_I) + \phi(3)\ \phi(4)\ \chi(\mathbf{R}_{II}) \right\} \tag{V,9}$$

where $\mathbf{R}_I \equiv \mathbf{R}_1 - \mathbf{R}_2$ and $\mathbf{R}_{II} \equiv \mathbf{R}_3 - \mathbf{R}_4$. In the case of deuteron $\alpha$-particle scattering $\phi(1), \phi(2), \phi(3), \phi(4)$ denote the internal functions of the *deuteron-*, $\alpha$-, *triton-* and $He^3$-cluster respectively. $\chi(\mathbf{R}_I)$ and $\chi(\mathbf{R}_{II})$ respectively describe the relative motion of the $\alpha$- and *deuteron*-clusters (channel I) and of the *triton-* and $He^3$-clusters (channel II). 1, 2, $\mathbf{R}_1, \mathbf{R}_2$ and 3, 4, $\mathbf{R}_3, \mathbf{R}_4$ represent the different internal and center of mass coordinates of the clusters and they are functions of the nucleon coordinates (in the deuteron $\alpha$-particle scattering example the coordinates $\mathbf{r}_1 \ldots \mathbf{r}_6$). For simplicity we again neglect at the moment all spin-orbit coupling effects. That means in the reaction process the total spin and also the total orbital angular momentum of the system are conserved independently of each other. The internal functions $\phi(1), \phi(2), \phi(3), \phi(4)$ are again considered as given functions which do not change their form during the whole reaction process, which again implies that we neglect any internal excitation or distortion of the clusters except exchange distortions arising from antisymmetrization*. This time we have two relative motion functions which are varied completely arbitrarily. If we

---

* Two clusters approaching one another at close range will tend to distend along their common line of centers due to the attractive nuclear forces, a polarization effect. Antisymmetrization also causes such polarization, see Fig. 10, 11 and 12.

introduce parameter coordinates and vary $\chi^*(R_I)$ and $\chi^*(R_{II})$ arbitrarily
in the expectation value of the Hamiltonian then we obtain completely
analogously to (V,6) the following coupled set of integro-differential
equations*

$$\left\{-\frac{\hbar^2}{2\,\mu_I}\,V^2_{R'_I} + V(R'_I)\right\} \chi(R'_I) + \int K(R'_I, R''_I)\,\chi(R''_I)\,dR''_I +$$
$$+ \int K(R'_I, R''_{II})\,\chi(R''_{II})\,dR''_{II} = E_I\,\chi(R''_I) \qquad\qquad (V,10)$$
$$\left\{-\frac{\hbar^2}{2\,\mu_{II}}\,V^2_{R'_{II}} + V(R'_{II})\right\} \chi(R'_{II}) + \int K(R'_{II}, R''_{II})\,\chi(R''_{II})\,dR''_{II} +$$
$$+ \int K(R'_{II}, R''_I)\,\chi(R''_I)\,dR''_I = E_{II}\,\chi(R''_{II})\;,$$

$\mu_I$ and $\mu_{II}$ are the reduced masses of the clusters *1* and *2* and of the
clusters *3* and *4* respectively, that is, they are the reduced masses of the
clusters in channels I and II. Analogously to above $V(R'_I)$ and $K(R'_I, R''_I)$
represent the local and nonlocal interaction terms of the clusters in
channel I, and similarly $V(R'_{II})$ and $K(R'_{II}, R''_{II})$ in channel II. $K(R'_I, R''_{II})$
and $K(R'_{II}, R''_I)$ are the transition kernels which produce transitions from
channel I to channel II and vice versa. $E_I = E_{TOT} - (E_{bind}(1) + E_{bind}(2))$
and $E_{II} = E_{TOT} - (E_{bind}(3) - E_{bind}(4))$ are the relative kinetic energies
of the clusters in channel I and II, or alternatively, the separation energies
in the case of closed channels. The constraint of normalization for the
wave function (V,9) during the variation we have taken into account by
introducing in the usual way a Lagrange multiplier equal to $-E_{TOT}$.
Again $V(R'_I)$ and $V(R'_{II})$ are real, and $K(R'_I, R''_I)$ and $K(R'_{II}, R''_{II})$ hermitian, and they are responsible for the elastic scattering of the particles in
channel I and channel II if one neglects the transition to the opposite
channel. Furthermore the transition kernels are also adjoint to one
another:

$$K(R'_I, R''_{II}) = K^*(R''_{II}, R'_I)\;. \qquad\qquad (V,11)$$

One sees this immediately from the definition of these kernels written
down before carrying out explicitly the integrations and summations
over the different variables:

$$K(R'_I, R''_{II}) = \iint \phi^*(1')\,\phi^*(2')\,\langle \mathscr{A}_r \delta(r_1 - r'_1) \cdots \delta(r_A - r'_A)\,|\,\bar{H}(r_i\,p_i) -$$
$$- E_{TOT}|\,\delta(r_1 - r''_1) \cdots \delta(r_A - r''_A)\rangle_\tau\,\phi(3'')\,\phi(4'')\,d\bar{\tau}'\,d\bar{\tau}''$$
$$= \iint \phi^*(1')\,\phi^*(2')\,\langle \delta(r_1 - r'_1) \cdots \delta(r_A - r'_A)\,|\,\bar{H}(r_i\,p_i) -$$
$$- E_{TOT}|\,\mathscr{A}_r \delta(r_1 - r''_1) \cdots \delta(r_A - r''_A)\rangle_\tau \cdot \phi(3'')\,\phi(4'')\,d\bar{\tau}'\,d\bar{\tau}'' \qquad (V,12)$$

and

$$K^*(R''_{II}, R'_I) = \iint \phi(3'')\,\phi(4'')\,\langle \mathscr{A}_r \delta(r_1 - r''_1) \cdots \delta(r_A - r''_A)\,|\,\bar{H}(r_i\,p_i) -$$
$$- E_{TOT}|\,\delta(r_1 - r'_1) \cdots \delta(r_A - r'_A)\rangle^*_\tau \cdot \phi^*(1')\,\phi(2')\,d\bar{\tau}'\,d\bar{\tau}''$$
$$= \iint \phi(3'')\,\phi(4'')\,\langle \delta(r_1 - r'_1) \cdots \delta(r_A - r'_A)\,|\,\bar{H}(r_i\,p_i) -$$
$$- E_{TOT}|\,\mathscr{A}_r \delta(r_1 - r''_1) \cdots \delta(r_A - r''_A)\rangle_\tau\,\phi(1')\,\phi^*(2')\,d\bar{\tau}'\,d\tau''\;.$$

---

* To fulfill $\delta\langle\bar{H}\rangle = 0$ one has to put both the coefficient of $\delta\chi^*(R_I)$ and $\delta\chi^*(R_{II})$
separately equal to zero.

In the matrix element $\langle|\bar{H} - E_{\text{TOT}}|\rangle_{\tau}$, which denotes sum and integration over the nucleon (unprimed) coordinates, we neglect integration over the (unprimed) center of mass coordinate because it is already split off. $\mathscr{A}_{r}$ antisymmetrizes just the nucleon coordinates, and to obtain the second expression for $K(\boldsymbol{R}'_{\text{I}}, \boldsymbol{R}''_{\text{II}})$ we use that $\mathscr{A}_{r}$ is hermitian and commutes with $(\bar{H} - E_{\text{TOT}})$. Again $d\bar{\tau}'$ and $d\bar{\tau}''$ denote that after introducing internal and relative center of mass cluster coordinates in the parameter coordinates one integrates and sums over all parameter coordinates except the coordinates $\boldsymbol{R}'_{\text{I}} = \boldsymbol{R}'_1 - \boldsymbol{R}'_2$ and $\boldsymbol{R}''_{\text{II}} = \boldsymbol{R}''_3 - \boldsymbol{R}''_4$. In the second expression for $K^*(\boldsymbol{R}''_{\text{II}}, \boldsymbol{R}'_{\text{I}})$ was used that $\bar{H}(\boldsymbol{r}_i \boldsymbol{p}_i)$ is a Hermitian operator.

We want to emphasize again that all integral kernels which appear in (V,10) are also functions of $E_{\text{TOT}}$. The just discussed properties of the direct interaction terms and the different integral kernels guarantee again the current conservation law, i.e. that the current of nucleons in the incoming particles is the same as the current of nucleons in the outgoing reaction products. They also guarantee (as in the one channel case) that solutions belonging to different energies are orthogonal.

Due to the rotational invariance of the Hamiltonian one can again, analogously to (V,8), expand the relative motion wave functions of the two cluster configurations in orbital angular momentum waves and obtain then for each orbital angular momentum value two coupled (just to the partial wave of the same $l$ value in the opposite channel) integro-differential equations, as can be seen easily. If we take into account spin-orbit coupling effects, the general structure of our coupled integro-differential equations is essentially unchanged excepting that the equations become more complicated due to the fact that one has to take into account the internal spins of the clusters explicitly; then the equations can be split, as in the first example, into coupled integro-differential equations to given total angular momenta, and further, to given parity.

If we apply our general formalism (V,10) to Li$^6$ as a specific example, then our equation can describe the following bound state configurations and reaction processes:

1. Bound state configurations of Li$^6$ and their energy eigenvalues which contain a superposition of an *α-deuteron* cluster structure and a *triton-He$^3$* cluster structure. Asymptotic condition: $\chi(\boldsymbol{R}'_{\text{I}})$ and $\chi(\boldsymbol{R}'_{\text{II}})$ must vanish exponentially for $\boldsymbol{R}'_{\text{I}} \to \infty$ and $\boldsymbol{R}'_{\text{II}} \to \infty$.

2. $\alpha$-deuteron elastic scattering and $\alpha$ (d, He$^3$) triton reaction ($E^{\text{CM}}_{\text{SC}} > 14.3 \text{ MeV}$). Asymptotic condition: In the $\alpha$-deuteron channel (I) superposition of an incoming and an outgoing wave for $\boldsymbol{R}_{\text{I}} \to \infty$. In the He$^3$-triton channel (II) only outgoing waves for $\boldsymbol{R}_{\text{II}} \to \infty$.

3. He$^3$-triton elastic scattering and He$^3$ (triton, $\alpha$) deuteron reaction. Asymptotic condition: In the He$^3$-triton channel (II) superposition of an incoming and an outgoing wave for $\boldsymbol{R}_{\text{II}} \to \infty$. In the $\alpha$-deuteron channel (I) only outgoing wave for $\boldsymbol{R}_{\text{I}} \to \infty$.

4. $\alpha$-deuteron elastic scattering ($E^{\text{CM}}_{\text{SC}} < 14.3 \text{ MeV}$) with inclusion of the bound *He$^3$-triton* cluster structure in the Li$^6$ compound nucleus formation. Asympotic condition: In the $\alpha$-deuteron channel super-

position of an incoming and an outgoing wave for $R_I \to \infty$. In the $He^3$-triton closed channel $\chi(R_{II})$ vanishes exponentially for $R_{II} \to \infty$. $E_{II} = E_{TOT} - (E_{bind}(3) + E_{bind}(4))$ is negative.

As discussed in chapter IV section A, particularly concerning the *α-deuteron* and *He³-triton* cluster configuration of Li⁶, the two cluster configurations of such two channel reactions are usually not orthogonal to each other, especially if one carries out in every channel the anti-symmetrization between the clusters. However the wave functions of the open channels are at least "asymptotically orthogonal" to each other. That means they are orthogonal to each other in a region where the separation distance of the clusters in each of the two channels is very large. In other words the clusters in the two channels do not overlap each other in the asymptotic region*. This is sufficient to define the experimentally measurable nuclear scattering and nuclear reactions cross sections uniquely in the usual way by means of the asymptotic wave functions in the different channels. Therefore in the here described method of coupled channel equations for the description of nuclear reaction and nuclear bound state problems one need not worry about orthogonality properties. Only if one calculates, for instance with bound state wave functions, gamma or β transition matrix elements, then one has to see that the bound state wave functions in the initial and final state are normalized properly as we have discussed earlier.

To illustrate further the flexibility of the here considered variational method for the derivation of equations with which one can treat nuclear structure and reaction problems in a unified way, we return to the example of elastic scattering of deuterons on α-particles. There we have discussed how one can use the coupled two channel equations to take into account the *He³-triton* cluster structure in the Li⁶ compund nucleus formation. (For $E_{SC}^{CM} < 14.3$ MeV in which we are interested here this is a closed channel.) One can simplify these calculations further by making a less refined approximation than the two channel approximation. For this one substitutes for the second term in (V,9) a given wave function for the description of the He³-triton configuration

$$\psi(He^3, t) = a \, \mathscr{A} \, \{\phi(3) \, \phi(4) \, F(R_{II})\} \tag{V,13}$$

where the relative motion function $F(R_{II})$ with $R_{II} = R_3 - R_4$ is now a given function, for instance of Gaussian form, and the amplitude $a$ is a variational parameter. Our ansatz is then:

$$\psi = \mathscr{A} \, \{\phi(1) \, \phi(2) \, \chi(R_I) + a \, \phi(3) \, \phi(4) \, F(R_{II})\} \,. \tag{V,13a}$$

If we vary now the expectation value of the energy, considering just $\chi(R_I)$ as free variational function and $a$ as variational parameter, then

---

* As pointed out before, at large separation distances the clusters behave like the corresponding free particles.

we obtain from $\delta \langle \overline{H} \rangle = 0$ the following equations[*]:

$$\left\{ -\frac{\hbar^2}{2\mu_{\mathrm{I}}} \nabla^2_{\mathbf{R}'_{\mathrm{I}}} + V(\mathbf{R}'_{\mathrm{I}}) \right\} \chi(\mathbf{R}'_{\mathrm{I}}) + \int K(\mathbf{R}'_{\mathrm{I}} \, \mathbf{R}''_{\mathrm{I}}) \, \chi(\mathbf{R}''_{\mathrm{I}}) \, \mathrm{d}\, \mathbf{R}''_{\mathrm{I}} + a \, G(\mathbf{R}'_{\mathrm{I}})$$

$$= E_{\mathrm{I}} \, \chi(\mathbf{R}'_{\mathrm{I}})$$

$$a \, B + \int F^*(\mathbf{R}'_{\mathrm{II}}) \, K(\mathbf{R}'_{\mathrm{II}}, \mathbf{R}''_{\mathrm{I}}) \, \chi(\mathbf{R}''_{\mathrm{I}}) \, \mathrm{d}\, \mathbf{R}''_{\mathrm{I}} \, \mathrm{d}\, \mathbf{R}'_{\mathrm{II}} = 0 \qquad \text{(V,14)}$$

with

$$G(\mathbf{R}'_{\mathrm{I}}) \equiv \int K(\mathbf{R}'_{\mathrm{I}}, \mathbf{R}''_{\mathrm{II}}) \, F(\mathbf{R}''_{\mathrm{II}}) \, \mathrm{d}\, \mathbf{R}''_{\mathrm{II}}$$

and

$$B \equiv \int [\mathscr{A} \, \phi^*(3) \, \phi^*(4) \, F^*(\mathbf{R}_{\mathrm{II}})] \, (\overline{H} - E_{\mathrm{TOT}}) \, \phi(3) \, \phi(4) \, F(\mathbf{R}_{\mathrm{II}}) \, \mathrm{d}\tau \, .$$

$G(\mathbf{R}'_{\mathrm{I}})$ is a given function of $\mathbf{R}'_{\mathrm{I}}$ and $B$ is a given constant. If we eliminate $a$ in the first equation by help of the second equation then we obtain again a one channel equation for the calculation of $\chi(\mathbf{R}'_{\mathrm{I}})$. Such an equation is much easier to evaluate on a computer than the two coupled integro-differential equations (V,10).

Equations of the form (V,14) are for instance useful for studying the influence of a resonance level on an elastic scattering process and for deriving the corresponding one level Breit-Wigner resonance formula. The Breit-Wigner formula is certainly contained in our approach. An example of this is the neutron-$\alpha$ scattering for a center of mass scattering energy around 17 MeV. The $\frac{3+}{2}$ level at 17.8 MeV can be described for this elastic scattering process approximately as a deuteron-triton bound state configuration even if the deuteron triton decay channel is just barely open at this energy. As an approximate deuteron-triton bound state wave function $\psi(t, d)$, one takes for instance the wave function (IV,5) where the variational parameters $a$, $b$ and $\beta$ are obtained by minimizing the energy expectation value there.

From our considerations of one channel and two channel reactions it is immediately clear how to take into account in a reaction process an arbitrary number of open and closed two cluster reaction channels. We simply employ as many terms from our mixed cluster expansion system as we care to treat. In fact one criterion for the completeness of any mixed cluster expansion system is that all channels be present. The set of coupled integro-differential equations which one obtains by help of the Ritz variational principle has then the following general form

$$\left\{ \frac{\hbar^2}{2\mu_N} \nabla^2_{\mathbf{R}'_N} + V(\mathbf{R}'_N) \right\} \chi(\mathbf{R}'_N) + \int K(\mathbf{R}'_N \, \mathbf{R}''_N) \, \chi(\mathbf{R}''_N) \, \mathrm{d}\, \mathbf{R}''_N +$$

$$+ \sum_{S \neq N} \int K(\mathbf{R}'_N \, \mathbf{R}''_S) \, \chi(\mathbf{R}''_S) \, \mathrm{d}\, \mathbf{R}''_S = E_N \, \chi(\mathbf{R}'_N) \, . \qquad \text{(V,15)}$$

In (V,15) we obtain a separate equation for each value of $N$, where $N$ goes over all open and closed channels considered. $S$ in (V,15) goes over all open and closed channels considered except the $N$-th channel. All kernels of (V,15) are again energy dependent. Further, the direct inter-

---

[*] The constraint of normalization we take into account again by introducing a Lagrange multiplier, $\lambda$.

action terms $V(\boldsymbol{R}'_N)$ are real and the kernels $K(\boldsymbol{R}'_N, \boldsymbol{R}''_N)$ are hermitian. For the transition kernels $K(\boldsymbol{R}'_N, \boldsymbol{R}''_S)$ we have analogous to (V,11) the adjoint relation $K(\boldsymbol{R}'_N, \boldsymbol{R}''_S) = K^*(\boldsymbol{R}''_S, \boldsymbol{R}'_N)$. As before, these properties of the different interaction terms guarantee the current conservation law for the incoming and outgoing particles and the orthogonality of solutions belonging to different energies, or in more high-brow words, these properties guarantee the unitarity of the $S$-matrix which one can derive from the coupled channel equations (V,15). $E_N$ represents again the relative kinetic energy, or in the case of closed channels the separation energy, of the two clusters in the $N$-th channel. If one were to take into account internally excited states of the clusters, then in the general form of the coupled equations (V,15) nothing would change because the new configurations would be described as additional closed or open channels.

By introducing more and more open and closed channels one can improve the description of any particular reaction process step by step. Certainly in practical calculations one can take into account only a very restricted number of channels. The channels which one has to take into consideration to come to a good approximate description of a reaction process will be selected usually from energetical reasons. Thus in low energy neutron-$\alpha$ scattering one can neglect in good approximation all other channels except the neutron-$\alpha$ channel, because due to the tight binding of the $\alpha$-cluster the transitions to other (closed) channels will be very weak. For neutron-$\alpha$ bombarding energies in the center of mass system around 17 MeV one has to take into consideration the just discussed deuteron-triton channel. If the neutron-$\alpha$ scattering energy increases further, then one has to take into consideration additional reaction channels. Other examples are $\alpha - \alpha$, $\alpha$-triton and $\alpha$-He$^3$ scattering, where for low scattering energies one can neglect the internal excitations of the clusters. With increasing scattering energies again more and more other reaction channels which describe successively the break-up of the *triton*- or *He$^3$*-clusters and $\alpha$-clusters and so on, play an important role, which means that the calculations become very complicated.

For high scattering energies one can very often make another approximation. If the transitions to the reaction channels are small compared with the current in the incoming channel then we can treat these transitions as perturbations on the dominating wave in the incoming channel. This is the first step in treating the coupled channel equations by an iteration method (Born approximation).

One can refine the calculations further by introducing additional variational parameters in the given internal functions of the clusters and in any given relative motion functions. If one does this in the variational principle then one obtains additional equations similar to the second equation (V,14) which come from the variation of these parameters*.

---

* The parameters of the internal cluster functions become functions of the cluster distances $\boldsymbol{R}_N$ and therefore would serve to take into consideration internal distortions of the clusters during the reaction processes.

In principle we can also use the variational method to derive equations for reactions where the compound nucleus disintegrates into 3 and more clusters. To do this one has to introduce in a cluster function two and more relative motion functions which can be varied freely. Thus if one wants to describe the $3\alpha$-particle decay of $C^{12}$ one has to introduce in the $C^{12}$-cluster function first a relative motion function which describes the relative motion of one $\alpha$-cluster against the other two $\alpha$-clusters and second a relative motion function which describes the relative motion of these other two $\alpha$-clusters against each other. To carry out the calculations for such multiple decay reactions explicitly becomes so complicated that even with the largest and fastest computers which exist today one can only treat at the moment the simplest reactions of this type.

Even without explicit calculation one can qualitatively understand with the cluster approach why multiple break-up decays occur sequentially rather than simultaneously if the decay levels of the residual nuclei are long lived, K. WATSON [62], A. LANE and R. THOMAS [62], [24], p. 114, G. PHILLIPS et al. [63], A. BAZ [64]. The lifetime of a level indicates how strongly the clusters mutually attract one another (see chapter IV section A and chapter V section D). If the attraction between clusters is strong (i. e. level width of the residual states less or of the order $\sim 1$ MeV), it is unlikely that the emission of one of the clusters will break apart the remaining clusters. Hence after the emission of one of the clusters the remaining clusters must be close to one another, which is only possible in the virtual states of the residual nucleus. It is thus understandable why for example the $3\alpha$-decay of many excited $C^{12}$ $(T = 0)$ levels proceeds mainly through the low virtual states of $Be^8$, and $3\alpha$-decay of these levels with a continuous energy spectrum is observed only with a small probability.

In conclusion we see from the considerations sketched above that the Ritz variational principle applied to cluster wave functions as variational functions is a very powerful and very flexible method to derive approximate equations for the description of nuclear reactions. This method also unifies the description of nuclear reactions and nuclear bound states in a very natural way. Even if one can carry out this kind of calculation explicitly at the moment only for relatively simple nuclear reactions and nuclear bound states one can use this method and the coupled integro-differential equations which one derives with it to discuss general properties of the nuclei and their reactions. For instance one sees immediately that the coupled channel equations method used in the phenomenological discussion of nuclear reactions follows quite naturally from our considerations. The different interaction terms are derived there phenomenologically from experimental data. One can understand therefore why this method when compared with other methods very often gives the best phenomenological description of nuclear reactions, see for instance R. DAVIS [66], and T. TAMURA [66].

The coupled channel equations which we have obtained by the Ritz variational principle applied on cluster representation wave functions can be considered also as generalized equations of the resonating group theory, J. WHEELER [67].

## C. Application to specific nuclear reactions

In this paragraph we will discuss quantitative and qualitative results obtained by applying the above described method to specific nuclear reactions.

### a) $\alpha - \alpha$ elastic scattering *

To describe the low energy elastic scattering of two $\alpha$-particles we make the following "ansatz" for the wave function:

$$\psi = \mathscr{A} \, \phi_0(\alpha_1) \, \phi_0(\alpha_2) \, \chi(\boldsymbol{R}) \qquad (V,16)$$

where $\chi(\boldsymbol{R})$ with $\boldsymbol{R} = \boldsymbol{R}_1 - \boldsymbol{R}_2$ is the variational function for the cluster relative motion, and $\phi_0(\alpha_1)$ and $\phi(\alpha_2)$ denote the given internal ground state functions of the two $\alpha$-clusters and their spin and isobaric spin configurations. The space part of $\phi_0(\alpha_1) \, \phi_0(\alpha_2)$ is as before assumed to be

$$\phi_0(\alpha_1) \, \phi_0(\alpha_2)_{space} = e^{-\frac{a}{2} \sum\limits_{i=1}^{4} (r_i - R_1)^2} \, e^{-\frac{a}{2} \sum\limits_{i=5}^{8} (r_i - R_2)^2} \qquad (V,16a)$$

with $a = 5.35 \cdot 10^{25}$ cm$^{-2}$, corresponding to a cluster "radius" of $1.45 \cdot 10^{-13}$ cm. This radius follows from the experimental root-mean-square charge radius $1.61 \cdot 10^{-13}$ cm in which account is taken of the finite charge distribution of the proton, reported to be $0.72 \cdot 10^{-13}$ cm for an assumed Gaussian charge distribution, R. HOFSTADTER [71].

A comparison with the treatment of Be$^8$ in the oscillator model (Chapter III Section C) shows that for an approximate description of Be$^8$ as a bound state one needs a value for the internal width parameter $a$ which is not very different from the above value of $a = 1.45 \cdot 10^{-13}$ cm. Therefore by choosing the $\alpha$-cluster radii equal to the experimental radius of the free $\alpha$-particle we take into account approximately the saturation character of the nuclear forces. Also by fixing $a$ we neglect any internal excitation and polarization of the $\alpha$-clusters ** during the scattering process.

Owing to the spherical symmetry of the $\alpha$-clusters, the spin-orbit and tensor forces will not contribute to $\alpha - \alpha$ scattering in a first approximation. Therefore we assume in the Hamiltonian a central two-nucleon potential of the form

$$V_{ik} = - V_0 \, e^{-\beta (r_i - r_k)^2} \left[ w - \frac{m}{4} (1 + \sigma_i \cdot \sigma_k)(1 + \tau_i \cdot \tau_k) \right.$$
$$\left. + \frac{b}{2} (1 + \sigma_i \cdot \sigma_k) + \frac{h}{2} (1 + \tau_i \cdot \tau_k) \right]. \qquad (V,17)$$

The parameters of our potential are chosen to be

$$V_0 = 72.98 \text{ MeV}; \quad \beta = 4.6 \cdot 10^{25} \text{ cm}^{-2}$$
$$w + m = b - h = 0.63; \quad w + m + b + h = 1. \qquad (V,17a)$$

---

* References: A. BUTCHER and J. MCNAMEE [68, 61]; E. SCHMID and K. WILDERMUTH [69], S. OKAI and S. PARK [70].
** Except exchange polarization (see chap. IV sect. D).

This potential yields a neutron-proton singlet effective range $r_{os}$, triplet effective range $r_{ot}$, neutron-proton scattering length $a_s$ and deuteron binding energy $E_d$ in close agreement with experiment, see for this [69], L. HULTHEN and M. SUGAWARA [72]. Also one obtains with this potential good agreement with the proton-proton scattering data, [69], [72]. The main part of our force mixture consists of a Serber mixture ($w = m$, $b = h$) which describes the symmetry of the two-nucleon scattering phase shifts around 90° in the center of mass system. The small deviation from this symmetry in the low energy region ($\leq 50$ MeV c.m.) we describe in first approximation by a small admixture of Rosenfeld force ($m = 2b$, $h = 2w$), which gives for instance the negative sign of the observed $P$-wave phase shifts for low energy p — p scattering, [72]. However these experimental data are not accurate enough to determine the amount of Rosenfeld force exactly. They only can tell that it is somewhere less than 10%. For the calculations reported here it is chosen to be 6%. We see that the two-nucleon force used here is approximately the same as the earlier used force (III,14).

For the determination of $\chi(\boldsymbol{R})$ we have corresponding to (V,8) the following integro-differential equation

$$\left\{-\frac{\hbar^2}{2\mu}\frac{\mathrm{d}^2}{\mathrm{d}R'^2} + \frac{\hbar^2 l(l+1)}{2\mu R'^2} + V(R')\right\} u_l(R') +$$

$$+ \int_0^\infty K_l(R', R'')\, u_l(R'')\, \mathrm{d}R'' = E\, u_l(R') \tag{V,18}$$

where $\dfrac{u_l(R')}{R'}$ is again the radial part of $\chi_l(\boldsymbol{R}')$ and $\mu = 2M_{\text{nucleon}}$ is the reduced cluster mass. In $V(R')$ also the Coulomb-interaction of the two clusters is contained. Due to the long range of the Coulomb forces the exchange part of the Coulomb interaction practically does not influence the scattering wave function $u_l(R')$, [70]. The explicit form of $V(R')$ and $K_l(R'\, R'')$ may be found in [61].

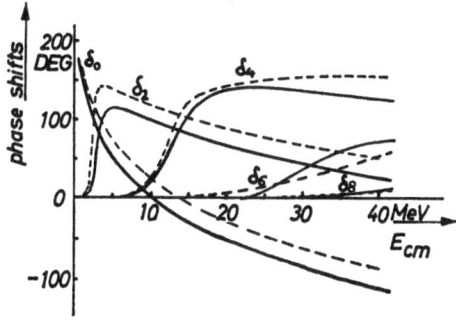

Fig. 14. Nuclear phase shifts vs. c.m. scattering energy for α — α elastic scattering; solid curve: theoretical, dotted curve: experimental

The results of the calculations for the angular momentum dependent phaseshifts $\delta_l$ as functions of the center of mass scattering energy are given together with the experimental phaseshifts in Fig. 14*. As one sees from Fig. 14 the theoretical and experimental phaseshifts agree well with each other. For a more detailed discussion of the results shown in Fig. 14 see [70]. There one finds also references to the experimental results.

### b) $\alpha - He^3$ elastic scattering

Completely analogously to $\alpha - \alpha$ elastic scattering one can describe $\alpha - He^3$ elastic scattering, Y. TANG et al. [73]. For the width parameters of the $\alpha$-cluster and the $He^3$-cluster one takes again the internal width parameters of the corresponding free particles. By fixing these parameters in the scattering process one takes into account again approximately the saturation character of the nuclear forces. The integro-differential equation for the radial part of the scattering wave function of orbital angular momentum $l$ has again the general form (V,8) and (V,18). The results of the calculations for two different center of mass scattering energies were obtained with the same forces used for the $\alpha - \alpha$ elastic scattering and are compared with the experimental data in Fig. 15 and 16.

In Fig. 15 and 16 this time the differential elastic scattering cross-sections are plotted; they are given by, see for instance [73],

$$\sigma(\theta) = |f(\theta)|^2 \tag{V,19}$$

with

$$f(\theta) = \frac{\eta}{2k \sin^2\left(\frac{\theta}{2}\right)} e^{-i\eta \ln \sin^2\left(\frac{\theta}{2}\right)} + \tag{V,19a}$$

$$+ \sum_{l=0}^{\infty} \frac{2\sqrt{\pi}}{k} \sqrt{2l+1} \; e^{i(2(\sigma_l - \sigma_0) + \delta_l)} \sin \delta_l \; Y_{l,0}(\theta) \; .$$

The notations in (V, 19) and (V, 19a) are the same as in (V,18a), except that $\mu$ which enters into $k$ is $\frac{12}{7} M_{\text{nucleon}}$**.

---

* The $\delta_l$ are here the phaseshifts relative to the Rutherford scattering phaseshifts. They are contained in the asymptotic form of $u_l$ in the following way, see for instance [68],

$$u_l(R' \to \infty) \sim \sin\left(kR' - \frac{1}{2} l\pi - \eta \ln 2kR + \sigma_l + \delta_l\right) \tag{V,18a}$$

where $\eta = \dfrac{4e^2}{\hbar v}$ with $v$ the relative velocity of the two $\alpha$-particles at infinity. The quantities $\sigma_l$ represent the Coulomb phaseshifts. The $\delta_l$ are the just mentioned nuclear phaseshifts relative to the Rutherford scattering phaseshifts. $k = \dfrac{\mu v}{\hbar}$

$= \dfrac{2Mv}{\hbar}$ is the wave number of the asymptotic relative motion.

** In the calculations spin-orbit forces were neglected. Therefore our calculations cannot provide for the spin-orbit splitting of the phaseshifts. Aside from this the agreement between theory and experiment looks quite good, especially if one takes into consideration that one scattering energy is relatively low (1.7 MeV) and the other relatively high (16.6 MeV).

We see from Fig. 16 that for 16.6 MeV scattering energy the calculated elastic differential cross-section is in average larger than the experimental cross-section. This is mainly due to the fact that we have not taken into account in our calculation the transitions to the p — Li$^6$ channel (threshold energy 4.02 MeV in the $\alpha$ — He$^3$ center of mass system [30]), as a rough estimation shows.

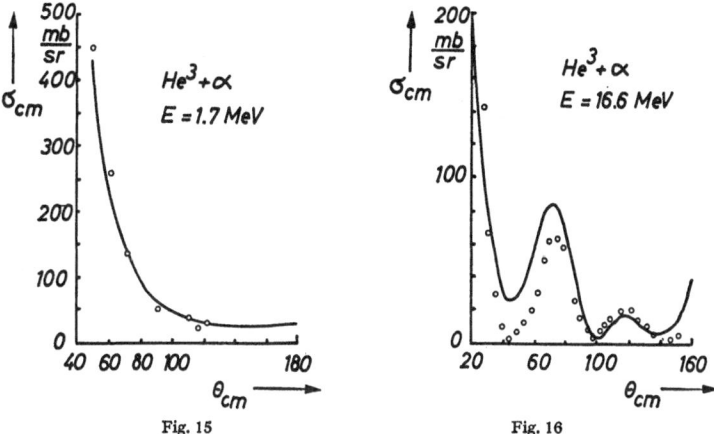

Fig. 15                        Fig. 16

Fig. 15. Angular distribution for $\alpha$ — He$^3$ elastic scattering at 1.7 MeV (c.m.); solid line; theoretical curve, circles: experimental points

Fig. 16. Angular distribution for $\alpha$ — He$^3$ elastic scattering at 16.6 MeV (c.m.); solid line: theoretical, circles: experimental points

Another result of the calculations is that for orbital angular momentum $l = 3$ a sharp resonant level appears at 5.0 MeV center of mass scattering energy. Experimentally the $^2f_{7/2}$ and the $^2f_{5/2}$ levels of Be$^7$, split by the spin-orbit coupling, occur at scattering energies of 4.9 and 2.8 MeV TOMBRELLO [168], from which we can deduce a mean resonance energy of 3.7 MeV. Compared to the experimental resonance energy the theoretical resonance energy is somewhat too high, which is again partly due to the fact that we have neglected the influence of other channels.

It turns out that the differential cross-sections as shown in Fig. 15 and 16 are rather insensitive against a change of the two-nucleon interaction potential as long as this potential describes the low energy two-nucleon scattering data well. For a more detailed discussion* see [73] where one finds also references to the experimental results. This supports the statement in the introduction that for the description of low energy nuclear phenomena it is only essential that we use a two-nucleon potential which describes correctly the low energy two-nucleon data.

The calculations discussed here show further that in the region where the clusters penetrate each other strongly the scattering wave functions agree roughly with the corresponding oscillator cluster model functions (which are equivalent to the usual oscillator shell model functions for these levels), as we except from the arguments in the first footnote of chap. IV sect. B. Further they show that the scattering wave

---

\* The same is true for the $\alpha$ — $\alpha$ scattering.

functions practically do not change in this region from 1.7 MeV to 16.6 MeV except for an energy dependent penetration factor (unpublished). The same is true for the $\alpha$ - $\alpha$ scattering wave functions[*] (unpublished).

### c) $Li^6(p, He^3)$ $\alpha$ reaction

As an example of a two channel reaction we discuss the $Li^6(p, He^3)\alpha$ reaction. Here we are especially interested in the question of how the description of nuclear reactions will be influenced by the fact that many nuclear states can be described almost equivalently by different simple cluster structures, E. SCHMID et al. [76]. For instance, as mentioned in chap. IV sect. A and shown in Appendix C the ground state of $Li^6$ can be described in the oscillator cluster model either by an $\alpha$-*deuteron* cluster structure or by a *triton-He$^3$* cluster structure. We will see that with such cluster representation equivalences the phenomena of light and heavy particle stripping and other reaction mechanisms are connected.

The incoming and outgoing wave in the proton-$Li^6$ elastic scattering channel we describe by the relative motion of a proton and a $Li^6$-cluster, whereas the $He^3$-$\alpha$ outgoing reaction channel involves relative motion of a $He^3$- and an $\alpha$-cluster. Our wave function must include at least these two cluster structures, and we approximate the wave function by assuming these two are the only cluster structures involved, i.e. we neglect transitions to all other reaction channels. As already shortly discussed in section B of this chapter this is reasonable provided the waves in the reaction channels are small enough to be treated as perturbations on the dominating wave in the proton-$Li^6$ channel, which condition usually holds in more or less good approximation at high scattering energies.

We restrict ourselves to the discussion of this first step of a Born approximation. In zeroth approximation we consider the scattering of the proton on $Li^6$ (channel I) without taking into consideration the transition to the $\alpha$-$He^3$ channel (channel II). Therefore we have analogously to (V,6) in zeroth approximation the following integro-differential-equation for the proton-$Li^6$ elastic scattering function

$$\left\{\frac{\hbar^2}{2\mu_I}\nabla_{\mathbf{R'_I}}^2 + V(\mathbf{R'_I})\right\}\chi_0(\mathbf{R'_I}) + \int K(\mathbf{R'_I}\,\mathbf{R''_I})\,\chi_0(\mathbf{R''_I})\,\mathrm{d}\mathbf{R''_I} = E_I\chi_0(\mathbf{R'_I}). \quad \text{(V,20)}$$

In first Born approximation we obtain then for the outgoing wave in the $\alpha$-$He^3$ channel the following inhomogenous integro-differential equation as can be seen immediately from (V,10),

$$\left\{\frac{\hbar^2}{2\mu_{II}}\nabla_{\mathbf{R'_{II}}}^2 + V(\mathbf{R'_{II}}) - E_{II}\right\}\chi_1(\mathbf{R'_{II}}) + \int K(\mathbf{R'_{II}},\mathbf{R''_{II}})\,\chi_1(\mathbf{R''_{II}})\,\mathrm{d}\mathbf{R''_{II}}$$
$$= -\int K(\mathbf{R'_{II}}\,\mathbf{R''_I})\,\chi_0(\mathbf{R''_I})\,\mathrm{d}\mathbf{R''_I}. \quad \text{(V,21)}$$

---

[*] The numerical calculations on the $\alpha$ — $\alpha$ scattering and the $\alpha$ — $He^3$ scattering were carried out at the IBM 7090 computer using methods given by L. Fox and E. GOODWIN [74] and by H. ROBERTSON [75].

To simplify our considerations further we assume that the proton-Li⁶ scattering energy is so high that we can neglect for the calculation of $\chi_0(\boldsymbol{R}_I')$ the distortion of the p-Li⁶ wave which comes from the local and nonlocal p-Li⁶ interaction. Similarly we assume that in the $\alpha$-He³ channel the relative energy of the $\alpha$- and He³-clusters is so large that we can also neglect the distortion of the $\alpha$-He³ wave by the local and nonlocal $\alpha$-He³ interaction. By these simplifications nothing is essentially changed in our general physical picture.

If we make these simplifications then the p-Li⁶ wave is represented by a plane wave of the form

$$\chi_0(\boldsymbol{R}_I') = \exp\{i\,\boldsymbol{k}_I \cdot (\boldsymbol{R}_p' - \boldsymbol{R}_{Li\bullet}')\} = \exp(i\,\boldsymbol{k}_I \cdot \boldsymbol{R}_I') . \qquad (V,22)$$

and we obtain for $\chi_1(\boldsymbol{R}_{II}')$ the following equation

$$\left\{-\frac{\hbar^2}{2\,\mu_{II}}\,\nabla^2_{R_{II}'} - E_{II}\right\} \chi_1(\boldsymbol{R}_{II}') = -\int K(\boldsymbol{R}_{II}',\boldsymbol{R}_I'')\,e^{i\,\boldsymbol{k}_I \cdot \boldsymbol{R}_I''}\,d\boldsymbol{R}_I'' \quad (V,23)$$

where from the general form of (V,12) we have

$$K(\boldsymbol{R}_{II}',\boldsymbol{R}_I'') = \int\!\!\int \phi^*(a')\,\phi^*(He^{3'})\,[\mathscr{A}_{r'}\delta(\boldsymbol{r}_1'' - \boldsymbol{r}_1') \cdots \delta(\boldsymbol{r}_7'' - \boldsymbol{r}_7')] \times$$
$$\times (\bar{H}(\boldsymbol{r}_i''\,\boldsymbol{p}_i'') - E_{TOT})\,\phi(Li^{6''})\,d\bar{\tau}'\,d\bar{\tau}'' \qquad (V,23a)$$

Because we are interested in the angular distribution of the reaction products we expand $\chi_1(\boldsymbol{R}_{II}')$ and $K(\boldsymbol{R}_{II}',\boldsymbol{R}_I'')$ in Fourier integrals as follows

$$\chi_1(\boldsymbol{R}_{II}') = \int a(\boldsymbol{k}_{II})\,e^{i\,\boldsymbol{k}_{II} \cdot \boldsymbol{R}_{II}'}\,d\boldsymbol{k}_{II} \qquad (V,24)$$

and

$$K(\boldsymbol{R}_{II}',\boldsymbol{R}_I'') = \int K(\boldsymbol{k}_{II},\boldsymbol{R}_I'')\,e^{i\,\boldsymbol{k}_{II} \cdot \boldsymbol{R}_{II}'}\,d\boldsymbol{k}_{II} \qquad (V,25)$$

with

$$K(\boldsymbol{k}_{II},\boldsymbol{R}_{II}') = \frac{1}{2\pi}\int \left\{e^{-i\,\boldsymbol{k}_{II} \cdot \boldsymbol{R}_{II}'}\,\phi^*(a')\,\phi(He^{3'}) \times \right.$$
$$\left. [\mathscr{A}_{r''}\delta(\boldsymbol{r}_1'' - \boldsymbol{r}_1') \cdots \delta(\boldsymbol{r}_7'' - \boldsymbol{r}_7')]\,(\bar{H}(\boldsymbol{r}_i''\,\boldsymbol{p}_i'') - E_{TOT})\,\phi(Li^{6''})\right\}d\bar{\tau}'\,d\bar{\tau}'' .$$
$$(V,25a)$$

Introducing (V,24), (V,25) and (V,25a) in (V,23) we obtain for $a(\boldsymbol{k}_{II})$

$$a(\boldsymbol{k}_{II}) = \frac{-1}{\dfrac{\hbar^2\,k_{II}^2}{2\,\mu_{II}} - (E_{II} + i\,\varepsilon)}\,\frac{1}{2\pi}\int \left\{e^{-i\,\boldsymbol{k}_{II} \cdot \boldsymbol{R}_{II}'}\,\phi^*(a')\,\phi^*(He^{3'}) \times \right.$$

$$\times [\mathscr{A}_{r'}\delta(\boldsymbol{r}_1'' - \boldsymbol{r}_1') \cdots \delta(\boldsymbol{r}_7'' - \boldsymbol{r}_7')]\left(\bar{H}\left(\boldsymbol{r}_i''\,\frac{\hbar}{i}\,\boldsymbol{V}_i''\right) - E_{TOT}\right)\phi(Li^{6''}) \times$$
$$\left. \times e^{-i\,\boldsymbol{k}_I \cdot \boldsymbol{R}_I''}\right\}d\bar{\tau}'\,d\bar{\tau}''$$

$$= -\frac{1}{2\pi}\left(\frac{\hbar^2\,k_{II}^2}{2\,\mu_{II}} - (E_{II} + i\,\varepsilon)\right)^{-1}\int [\mathscr{A}_{r'}\,\phi^*(a')\,\phi^*(He^{3'})\,e^{-i\,\boldsymbol{k}_{II} \cdot \boldsymbol{R}_{II}'}] \times$$
$$\times (\bar{H} - E_{TOT})\,\phi(Li^{6'})\,e^{i\,\boldsymbol{k}_I \cdot \boldsymbol{R}_I'}\,d\bar{\tau}'$$

$$\equiv -\frac{1}{2\pi}\left(\frac{\hbar^2\,k_{II}^2}{2\,\mu_{II}} - E_{II} - i\,\varepsilon\right)^{-1}B(|\boldsymbol{k}_I|, |\boldsymbol{k}_{II}|, \cos\gamma) \qquad (V,26)$$

In the second expression of (V,26) one has to express the coordinates $(R'_{II})$, $(a')$, $(He^{3'})$ as functions of the coordinates $r'_1 \cdots r'_7$ and then to antisymmetrize. The factor $\left(\frac{\hbar^2 k_{II}^2}{2\,\mu_{II}} - (E_{II} + i\,\varepsilon)\right)^{-1}$ serves to insure first energy conservation and second, due to the infinitesimal positive imaginary addition $i\,\varepsilon$, that we have only outgoing $\alpha$-He$^3$ waves in the $x$-space.

The quantity $B$ appearing in (V,26)

$$B \equiv \int [\mathscr{A}\,\phi^*(a)\,\phi^*(He^3)\,e^{-i\,k_{II}\cdot R_{II}}]\,(\bar{H} - E_{TOT})\,[\phi(Li^6)\,e^{i\,k_I\cdot R_I}]\,d\tau$$

$$(V,26a)$$

is the reaction amplitude. $|B|^2$ is proportional to the probability as to which direction relative to the incoming proton beam the $\alpha$ and He$^3$ particles are emitted. We see that $B$ depends only on the angle $\gamma$ between $\hbar\,k_I$ and $\hbar\,k_{II}$, the relative momenta of the incoming beam and target and the outgoing reaction products respectively, because we describe the $a$, $He^3$ and the $Li^6$ ground state clusters with spherically symmetric wave functions and neglect spin-orbit coupling. We shall now discuss the angular dependence of $B$ in more detail.

In the limit case considered here where the incoming protons and the emitted reaction products have high kinetic energies (compared with the mean kinetic energy of the nucleons in Li$^6$) the factor $\exp\{i(k_I\cdot R_I - k_{II}\cdot R_{II})\}$ and its exchange terms will oscillate very fast as functions of $R_I$ and $R_{II}$ and therefore will have a decisive influence on the magnitude of the integral which defines $B$ as a function of the angle $\gamma$. If this factor oscillates especially rapidly the negative and positive contributions to the integral tend to cancel and the value of the integral becomes small; if the oscillation is less fast then the integral value tends to be larger. Further, due to our assumptions about the internal structure of the $Li^6$-, $a$-, and $He^3$-clusters, from this factor and its exchange terms comes the complete $\gamma$ dependence of $B$. Hence we must investigate the behaviour of this factor and its exchange terms as functions of the angle $\gamma$.

For the unsymmetrized initial state (right hand side) in the matrix element defining $B$ in (V,26a) let us arbitrarily assume that nucleons 1 to 6 build up the $Li^6$-cluster and the incoming proton is nucleon 7. For the (antisymmetrized) final state in $B$ let us assume for the direct term that nucleons 1 to 4 comprise the $a$-cluster and nucleons 5 to 7 the $He^3$-cluster. Thus for the direct term in $B$ we have the following definitions of $R_I$ and $R_{II}$

$$R_I = r_7 - \frac{1}{6}\,(r_1 + r_2 + r_3 + r_4 + r_5 + r_6)$$

$$R_{II} = \frac{1}{3}\,(r_5 + r_6 + r_7) - \frac{1}{4}\,(r_1 + r_2 + r_3 + r_4)\,.$$

$$(V,27)$$

For simplicity we assume that the mass of Li$^6$ is very large compared with the proton mass, i.e. we neglect in (V,27) the center of mass coordinate of Li$^6$. This simplifies the discussion without changing anything essential. In the direct term of $B$ the factor $\exp\{i(k_I\cdot R_I - k_{II}\cdot R_{II})\}$

obtains therefore the following form★

$$\exp\left\{i\left(k_{\mathrm{I}}-\frac{1}{3}k_{\mathrm{II}}\right)\cdot r_7\right\}\exp\left\{-ik_{\mathrm{II}}\cdot\left(\frac{1}{3}(r_5+r_6)-\frac{1}{4}(r_1+r_2+r_3+r_4)\right)\right\}$$

$$(V,27a)$$

where only the first factor depends on $k_{\mathrm{I}}$ and therefore on the direction between $k_{\mathrm{I}}$ and $k_{\mathrm{II}}$. Clearly $\frac{1}{3}|k_{\mathrm{II}}|$ is, at least in the high scattering energy limit case, smaller than $|k_{\mathrm{I}}|$. Therefore the oscillation of (V,27a) as a function of $r_7$ will be slowest if $k_{\mathrm{II}}$ is parallel to $k_{\mathrm{I}}$, which corresponds to a minimum momentum change of the incoming proton. Therefore the direct term of $B$ will have a maximum about this direction, which corresponds to the He³ particle coming off in the forward direction parallel to the incoming proton. The same is true for all exchange terms $B$ in which, in the final state, nucleon 7 is not permuted out of the $He^3$-cluster.

On the other hand all exchange terms of $B$ in which, in the final state, nucleon 7 is permuted into the $\alpha$-cluster will be small for $k_{\mathrm{I}}$ parallel $k_{\mathrm{II}}$ because now the factor depending on $\gamma$ which oscillates as a function of $r^7$ looks like

$$\exp\left\{i\left(k_{\mathrm{I}}+\frac{1}{4}k_{\mathrm{II}}\right)\cdot r_7\right\}$$

$$(V,27b)$$

and will oscillate very fast. In fact these exchange terms will be a maximum when $k_{\mathrm{II}}$ is anti-parallel to $k_{\mathrm{I}}$, which corresponds to the $\alpha$-particle coming off in the forward direction (and, since it contains nucleon 7, to a minimum momentum change of the incoming proton again) and the He³ particle in the backward direction★★. Thus we can say those terms of $B$ in which the incoming proton 7 ends up in the $He^3$-cluster of the final state contribute a forward peak to the He³ reaction cross-section, and terms of $B$ in which the incoming proton ends up in the final state $\alpha$-cluster contribute a backward peak to the He³ cross-section. With increasing scattering energy of the proton $|k_{\mathrm{I}}|$ and $|k_{\mathrm{II}}|$ will increase and the just discussed oscillation will cause the integrals defining $B$ to peak more sharply as a function of the angle $\gamma$. Certainly the absolute value of $B$ will go down with increasing scattering energy, but the peaks at zero and 180 degrees will become narrower in angular spread.

The above qualitative remarks take into account the complete anti-symmetrization of the nuclear wave function and show that for high bombarding energies we may expect a reasonably sharp forward and backward peak in the He³ reaction cross-section. Let us now try to

---

★ For simplicity we ignore any space exchange terms in the nuclear potential. Our arguments are not essentially altered by this simplification.

★★ Li⁶ is represented best by an $\alpha-d$ cluster structure. Note however this means that if the $\alpha$-cluster goes off in the forward direction (backward He³ peak) we must replace one of the original $\alpha$-cluster nucleons by the incoming nucleon 7. This exchange reduces the backward peak when calculated with generalized cluster functions compared to a calculation using the oscillator assumption for Li⁶, because the $\alpha$- and $d$-cluster are more highly overlapped in oscillator cluster functions.

understand these two peaks in the language of direct reaction theory. In this theory only partial antisymmetrization is performed. In the initial state antisymmetrization between the incoming nucleon and the target is neglected; only the target itself is antisymmetrized. In the final state the incoming nucleon is presumed to always end up in one given cluster and any permutations which remove the proton from that cluster are neglected. We shall see that because many nuclei can be almost equivalently described by several different cluster structures, this failure to completely antisymmetrize often makes the designation of the reaction mechanism (stripping, knock-out, etc.) rather equivocal.

To clarify these remarks let us reconsider our example of Li⁶ bombarded by a fast proton in the framework of the above direct reaction theory. Suppose we wish to describe the He³ reaction as a deuteron pick-up process. In this picture we consider the Li⁶ target to have an $a - d$ cluster structure, and the incoming proton picks up the *deuteron* cluster and goes off as a He³ particle. Hence in the final state we may only consider such exchange terms in which the incoming nucleon 7 remains in the $He^3$-cluster. Hence we write the deuteron pick-up amplitude as

$$B_{\mathrm{dp}} \cong \langle \mathscr{A}' \phi_\alpha(1\ 2\ 3\ 4)\ \phi_{\mathrm{He}^3}(5\ 6\ 7) \exp\{i\,\boldsymbol{k}_{\mathrm{II}}\cdot\boldsymbol{R}_{\mathrm{II}}\}\,|\bar{H} - E_{\mathrm{TOT}}|$$
$$\{\mathscr{A}_6 \phi_\alpha(1\ 2\ 3\ 4)\ \phi_\mathrm{d}(5\ 6)\ \chi(a - d)\} \exp\{i\,\boldsymbol{k}_{\mathrm{I}}\cdot\boldsymbol{r}_7\}\rangle$$
$$= 6!\,\langle \mathscr{A}' \phi_\alpha(1\ 2\ 3\ 4)\ \phi_{\mathrm{He}^3}(5\ 6\ 7) \exp\{i\,\boldsymbol{k}_{\mathrm{II}}\cdot\boldsymbol{R}_{\mathrm{II}}\}\,|\bar{H} - E_{\mathrm{TOT}}|$$
$$\{\phi_\alpha(1\ 2\ 3\ 4)\ \phi_\mathrm{d}(5\ 6)\ \chi(a - d)\} \exp\{i\,\boldsymbol{k}_{\mathrm{I}}\cdot\boldsymbol{r}_7\}\rangle \qquad (\mathrm{V,27c})$$

where $\mathscr{A}_6$ completely antisymmetrizes nucleons 1 to 6 (i.e. the Li⁶ target) and $\mathscr{A}' \equiv \mathscr{A}_6(1 - P_{57} - P_{67})$ includes all permutations of the final state in which the incoming nucleon remains in the $He^3$-cluster*. But we see that $B_{\mathrm{dp}}$ includes exactly those terms which by our earlier discussion contribute to the forward peak in the He³ reaction cross-section. Since in our earlier discussion the exchange terms neglected here (having nucleon 7 in the $a$-cluster) contribute very little to the forward peak, the above reaction amplitude $B_{\mathrm{dp}}$ will describe this forward peak quite well at high incident energy of the proton; it will of course give no backward peaking. Hence we may think of the forward peak as arising from a deuteron pick-up process.

However, we should recall that in the oscillator cluster model the antisymmetrized $a - d$ cluster description of Li⁶ is equivalent to the antisymmetrized $t - He^3$ cluster description (Appendix C). If we use generalized cluster functions these two descriptions still remain almost equivalent, which means to a good approximation we can substitute

$$\mathscr{A}_6\{\phi_\alpha(1\ 2\ 3\ 4)\ \phi_\mathrm{d}(5\ 6)\ \chi(\alpha \cdot d)\} \text{ by } \mathscr{A}_6\{\phi_{\mathrm{He}^3}(1\ 2\ 3)\ \phi_t(4\ 5\ 6)\ \chi(He^3 - t)\}$$

in $B_{\mathrm{dp}}$, whereupon we have

$$B_{\mathrm{dp}} = 6!\,\langle \mathscr{A}' \phi_\alpha(1\ 2\ 3\ 4)\ \phi_{\mathrm{He}^3}(5\ 6\ 7) \exp\{i\,\boldsymbol{k}_{\mathrm{II}}\cdot\boldsymbol{R}_{\mathrm{II}}\}\,|\bar{H} - E_{\mathrm{TOT}}|$$
$$\{\phi_{\mathrm{He}^3}(1\ 2\ 3)\ \phi_t(4\ 5\ 6)\ \chi(He^3 - t)\} \exp\{i\,\boldsymbol{k}_{\mathrm{I}}\cdot\boldsymbol{r}_7\}\rangle \qquad (\mathrm{V,27d})$$

---

* In the above notation $P_{57}$ and $P_{67}$ are pair exchanges; thus $P_{57}$ exchanges nucleons 5 and 7.

Since this is the same matrix element as before it again has a forward peak in the He$^3$ cross-section, corresponding to minimum change in the momentum of the incoming proton. However the mechanism here can be termed a He$^3$ knock-out process, for the proton is incident upon a *triton* plus *He$^3$*-cluster and a *He$^3$*-cluster comes off in the forward direction. Thus we may think of the forward peak more or less arbitrarily as either a deuteron pick-up or a He$^3$ knock-out, depending on which almost equivalent cluster representation is chosen to describe the ground state of the Li$^6$ target.

Let us now go further and imagine the He$^3$ reaction as a triton pick-up process. In this picture we again consider the Li$^6$ target to have a $t - He^3$ cluster structure, and the incident proton picks up the triton and goes out as an $\alpha$-particle leaving the He$^3$ particle to recoil. Hence in the final state we may only consider such exchange terms which leave the incoming proton 7 in the $\alpha$-cluster. Hence we have for the triton pick-up reaction amplitude:

$$B_{tp} = \langle \mathscr{A}'' \phi_{He^3}(1\ 2\ 3)\ \phi_\alpha(4\ 5\ 6\ 7)\ \exp\{i\ k_{II}\cdot R_{II}\}$$
$$|\bar{H} - E_{TOT}|\ \{\mathscr{A}_6 \phi_t(1\ 2\ 3)\ \phi_{He^3}(4\ 5\ 6)\ \chi(t - He^3)\}\cdot \exp\{i\ k_I r_7\}\rangle$$
$$= 6!\ \langle \mathscr{A}'' \phi_{He^3}(1\ 2\ 3)\ \phi_\alpha(4\ 5\ 6\ 7)\ \exp\{i\ k_{II}\cdot R_{II}\}$$
$$|\bar{H} - E_{TOT}|\ \{\phi_t(1\ 2\ 3)\ \phi_{He^3}(4\ 5\ 6)\ \chi(t - He^3)\}\cdot \exp\{i\ k_I r_7\}\rangle \quad \text{(V,27e)}$$

where $\mathscr{A}_6$ as before antisymmetrizes nucleons 1 to 6, but here $\mathscr{A}'' \equiv \mathscr{A}_6 \times$ $\times (1 - P_{47} - P_{57} - P_{67})$ goes over all permutations of the final state in which nucleon 7 remains in the $\alpha$-cluster. By our earlier discussion we see that $B_{tp}$ includes only those terms which contribute strongly to the backward peak in the He$^3$ reaction cross-section, corresponding to a forward peak in the $\alpha$-particle cross-section which means the incoming proton alters its relative motion as little as possible. Since in the earlier completely antisymmetrized treatment the exchange terms neglected in $B_{tp}$ here contribute very little to the backward peak, then $B_{tp}$ will describe the backward peak very well. Hence we may think of the backward peak as arising from a triton pick-up process.

However if we again substitute equivalent cluster descriptions we may also write $B_{tp}$ as

$$B_{tp} = 6!\ \langle \mathscr{A}'' \phi_{He^3}(1\ 2\ 3)\ \phi_\alpha(4\ 5\ 6\ 7)\ \exp\{i\ k_{II}\cdot R_{II}\}$$
$$|\bar{H} - E_{TOT}|\ \{\phi_\alpha(1\ 2\ 3\ 4)\ \phi_d(5\ 6)\ \chi(\alpha - d)\}\cdot \exp\{i\ k_I\cdot r_7\}\rangle \quad \text{(V,27f)}$$

$B_{tp}$ is in good approximation mathematically unaltered by this substitution* so that it again accounts for the backward peak of He$^3$ particles. But pictorially we now have the proton incident upon an $\alpha$- and a $d$-cluster and an $\alpha$-cluster leaves in the forward direction. Hence we may equivalently describe the backward peak in the He$^3$ cross-section as an $\alpha$ knock-out process as well as a triton pick-up.

---

* In changing from one cluster description to an equivalent cluster description we do not worry about normalization questions, which are not essential for our qualitative considerations.

To sum up, in the direct reaction treatment if we describe Li⁶ with an $a - d$ cluster structure then the forward peak of the He³ reaction cross-section may be ascribed to a deuteron pick-up and the backward peak to an $\alpha$ knock-out. If we describe Li⁶ by the almost equivalent $t - He^3$ cluster structure, we can ascribe the forward peak to He³ knock-out and the backward peak to triton pick-up. Alternatively, if we stick to pick-up processes, we may say that for He³ produced in the forward direction Li⁶ "looks like" an $\alpha$-cluster plus a $d$-cluster, whereas in the backward direction Li⁶ "looks like" a $He^3$-cluster plus a $t$-cluster. In the completely antisymmetrized treatment both peaks come out automatically and these rather arbitrary designations do not arise.

Similar considerations can be made also for many other reactions. For instance, for the F¹⁹ (p, $\alpha$) reaction at high scattering energies one can conclude completely analogously that for forward scattering of the outgoing $\alpha$-particles F¹⁹ looks like O¹⁶ plus triton, and for the backward scattering like N¹⁵ plus an $\alpha$, see for instance S. Edwards [77]. In the oscillator model F¹⁹ can be represented as

$$\psi(F^{19}) = \mathscr{A} \, \phi(\underset{\sim}{t}) \, \phi(\underset{\sim}{O^{16}}) \, \chi(t - O^{16})$$
$$= N \mathscr{A} \, \phi(\underset{\sim}{\alpha}) \, \phi(\underset{\sim}{N^{15}}) \, \chi(\alpha - N^{15})$$

where $N$ is a normalization constant. $\chi(t - O^{16})$ is a harmonic oscillator function with six excitation quanta and $\chi(\alpha - N^{15})$ is a harmonic oscillator function with seven quanta of excitation. In some respect the F¹⁹(p, $\alpha$) reaction is a much better example of the special direct reaction mechanisms discussed above, because the deviation of F¹⁹ from the oscillator assumption is relatively small*.

To carry out the above sketched calculations in detail becomes already rather complicated for the Li⁶ (p,He³) $\alpha$ reaction, especially if one wants to solve the belonging coupled set of integro-differential equations more exactly. Therefore a one dimensional two-channel problem with a target nucleus consisting of 4 nucleons which can be described by two different cluster structures was investigated in great detail, S. Okai et al. [78]**. This example was constructed in such a way that no essential features of the nuclear reaction mechanism, especially direct reaction mechanisms, were lost. In the investigation of this example it turned out, as our above qualitative considerations implied, that the usual designations of direct reactions as stripping reactions, pick-up reactions, knock-out reactions, etc., is not unequivocal. These different reaction mechanisms overlap each other sometimes very strongly. As above, it turned out for instance that the designation of a direct reaction process as a pick-up reaction or knock-out reaction, and so on, depends completely on the

---

* For these and similar reactions the so-called direct reaction mechanism which explains the forward peak of the light reaction products is very often called light particle pickup or stripping (i. e. in F¹⁹ the incoming proton picks up a *triton* cluster and goes forward as an $\alpha$) and that which explains the forward peak of heavy reaction products is called heavy particle pick-up or stripping (i. e. in F¹⁹ the incoming proton picks up an N¹⁵-cluster and goes off as O¹⁶).

** Reference [78] further elaborates the considerations sketched here.

system of cluster wave functions which one chooses for the description of the target nucleus. Another important result of this investigation was that direct reactions are essentially surface reactions. But this is intimately connected with the Pauli principle and not so much with the absorption of the impinging particle in the target nucleus as usually assumed.

By solving the coupled equations describing this two channel problem with two different methods – a matrix method and an iteration method[*] – it was also possible to study the validity of the usually applied phenomenological direct reaction calculations. In this connection one could explicitly see why at high scattering energies the distorted wave Born approximation describes the reaction cross-sections much better than the plane wave Born approximation, which always gives much too large cross-sections.

## D. Optical potential and nuclear reactions

In the phenomenological description of nuclear scattering and reaction processes the introduction of an optical potential for the description of the interaction and absorption (complex optical potential) of the reacting particles is very useful, see for instance [79]. We shall discuss how such optical potentials can be derived from our coupled channel equations at least in principle.

### a) Real optical potentials

We start with the discussion of optical potentials of purely elastic scattering processes, i.e. of scattering processes where one has no transitions to other channels or neglects them. At first we consider the elastic scattering of two clusters. We write down the general form of the integro-differential equation for given orbital angular momentum $l$ which describes the elastic scattering[**]

$$\left\{-\frac{\hbar^2}{2\mu}\frac{d^2}{dR'^2} + \frac{\hbar\, l(l+1)}{2\mu R'^2} + V(R') - E\right\} u_l(R')$$

$$= -\int_0^\infty K_l(R', R'')\, u_l(R'')\, dR'' \qquad (V,28)$$

or

$$\left\{\frac{\hbar^2}{2\mu}\frac{d^2}{dR'^2} + E\right\} u_l(R') = \int_0^\infty \overline{K}_l(R', R'')\, u_l(R'')\, dR'', \qquad (V,29)$$

if we include the centrifugal term $\dfrac{\hbar^2\, l(l+1)}{2\mu R'^2}$ and the local interaction term $V(R')$ of the clusters in the kernel $\overline{K}_l(R', R'')$.

---

[*] Iterated Born approximation.
[**] We neglect again spin-orbit coupling effects.

The optical potential to the scattering wave function $u_l(R')$ is defined as an energy and orbital angular momentum dependent real potential which is diagonal in the $x$-space and produces the same scattering wave as the integral kernel in (V,29). It is therefore defined by the following equation, E. SCHMID et al. [80]:

$$\left\{\frac{\hbar^2}{2\mu}\frac{d^2}{dR'^2} + E\right\} u_l(R') = V_{opt}(E, l, R')\, u_l(R')\,, \qquad (V,30)$$

$$V_{opt}(E, l, R') \equiv \frac{\left\{\dfrac{\hbar^2}{2\mu}\dfrac{d^2}{dR'^2} + E\right\} u_l(R')}{u_l(R')}\,. \qquad (V,30a)$$

In Fig. 17 the optical potential of $\alpha - He^3$ elastic scattering for $l = 0$ at two different scattering energies ($E = 1.7$ MeV and $E = 16.6$ MdV) is given, [80].

In the calculation of the optical potential the Coulomb interaction of the clusters was neglected, as is usually done in phenomenological scattering theories. One sees from Fig. 17 that despite the large energy spread between the two considered scattering energies the optical potential of the $\alpha - He^3$ clusters does not change very much from low energy scattering (1.7 MeV c.m. energy) to high energy scattering (16.6 MeV c.m. energy). This agrees with the usual behaviour of optical potentials in phenomenological scattering theories. The $\alpha - He^3$ optical potential also behaves smoothly through the $\alpha - He^3$ resonance energy region around 5 MeV in the c.m. frame*.

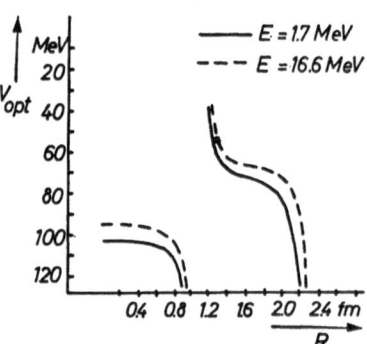

Fig. 17. Optical potential for $l = 0$ elastic $\alpha - He^3$ scattering

We want to mention another feature of optical potentials of the form (V,30) and (V,30a), [80]. As we see from Fig. 17, such potentials have singularities as functions of $R'$. This arises from the fact that a wave function obtained from a nonlocal potential can have non-zero second derivative while the function itself is equal to zero. It can be shown that no singularities can occur in nuclear matter because of the translation invariance of the interaction, but they seem to be a general feature of the nuclear surface region where the wavelength of the relative motion changes rapidly. This follows from generalization of our $\alpha - He^3$ considerations. There will be no singularities if an absorption is present**, but the optical potential will still have a rather rapidly varying shape at the nuclear surface. This seems to indicate that in optical model calculations, especially in distorted wave calculations in the surface region the optical

---

* The $\alpha - He^3$ optical potential would only change suddenly in the resonance energy region if another cluster structure would be responsible for the resonance behaviour, H. BENOEHR and K. WILDERMUTH [81].

** i. e. if there are transitions to other channels, as will be discussed below.

potential must be sometimes chosen rather rapidly varying to describe the experimental scattering data.

In chapter IV section A it was pointed out that the width of resonance levels depends primarily on the effective interaction between the principal clusters forming these levels. But as already mentioned, the argument there was not completely correct because the potential barrier which we have regarded as a measure for the interaction between the clusters depends on the bound state form of the wave packets with which we described the relative motion of the clusters. On the other hand resonance levels are usually excited by scattering processes, for instance $\alpha$-neutron scattering or deuteron-triton scattering as in our He⁵ example, and the relative motions of the clusters must then be described by scattering waves and not by wave packets of Gaussian form. Therefore one has to look for a quantity which is a measure of the interaction strength between the clusters in such scattering processes. The most natural quantities in this respect are, as already discussed, the corresponding energy and angular momentum dependent optical potentials which can be calculated from equations of the form (V,18), (V,29) and (V,30a), E. HOHLOCH and K. WILDERMUTH [82].

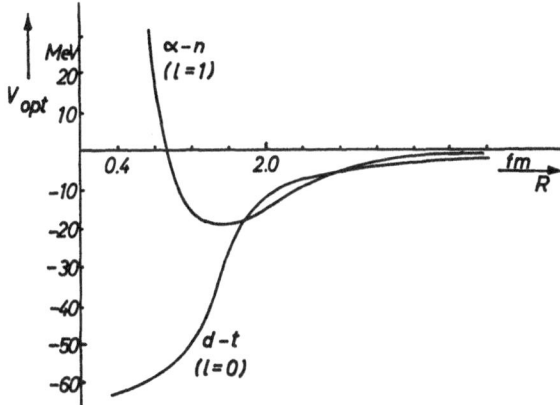

Fig. 18. Optical potential for $\alpha$ — n scattering ($l = 1$) and t — d scattering ($l = 0$) at 1 MeV

For example, in Fig. 18 are given the optical potentials of neutron-$\alpha$ scattering for $l = 1$ and deuteron-triton scattering for $l = 0$, [82]. In both cases the scattering energy in the center of mass system is 1 MeV. For the calculation of $K_l(R', R'')$ and $\overline{K}_l(R', R'')$ again the two-body forces (V,17) were used. To take into account approximately the saturation character of the nuclear forces the internal width parameters for the different clusters are again set equal to the internal width parameters of the corresponding free particles.

One sees from Fig. 18 that the deuteron-triton optical potential is deeper and also wider than the neutron-$\alpha$ optical potential. This indicates that the effective interaction between the *triton-* and the *deuteron-* cluster is much stronger than that between the neutron and the $\alpha$-cluster.

Therefore it is immediately understandable why in He[5] the $\frac{3^-}{2}$ level width (excitation energy 0.95 MeV) is much larger than the $\frac{3^+}{2}$ level width (excitation energy 16.7 MeV*.) Another factor which favours this tendency is that the reduced mass of the *deuteron-triton* cluster system is larger by a factor 1.5 than the reduced mass of the *neutron-$\alpha$* cluster system. The quantitative calculation of the two scattering cross-sections as functions of the scattering energies confirms these conclusions on the relative level widths.

In both cases considered above, the optical potentials change again relatively slowly with the scattering energy. They remain practically unchanged for scattering energies between 0 an 1 MeV although in both cases resonance levels lie in this scattering energy range. It should be mentioned that in the calculation of the optical potentials spin-orbit forces and Coulomb forces were neglected again. Ignoring such terms has no essential influence on our general considerations. Hence one sees from the above remarks that the considerations in chapter IV section A about the energy width of resonance levels remain valid if the cluster interaction energies defined there are replaced by the appropriate optical potentials.

If one wants to calculate the optical potential for the elastic scattering of light clusters on heavy clusters then the calculations become practically unmanageable. Only in the limit case where one neglects nuclear surface effects can these calculations be carried out. The results therefore are applicable only to target nuclei having large mass number $A$, i.e. approximately nuclear matter. For applications to finite nuclei correction terms have to be used. We shall sketch very shortly these calculations and their results, H. WITTERN [83].

For the wave function of the target and incident cluster we make the following ansatz:

$$\psi = \mathscr{A}\left\{\phi_{\text{cl}}\phi_{\text{tar}}\chi(R_{\text{cl}} - R_{\text{tar}})\right\}. \tag{V,31}$$

The functions $\phi_{\text{cl}}$ and $\phi_{\text{tar}}$ are considered as given. To comply approximately with the nuclear saturation condition the internal width parameter of $\phi_{\text{cl}}$, which is described by means of oscillator wave functions, is again set equal to the width parameter of the free cluster; and the density of $\phi_{\text{tar}}$, which later is approximated by a nuclear matter wave function $\phi_{\text{n.m.}}$, is chosen equal to the nucleon density in heavy nuclei. From the Ritz variational principle and using our usual two-body forces (III,14) or (V,17 and 17a) we obtain an integro-differential equation for $\chi(R_{\text{cl}} - R_{\text{tar}})$ of the form:

$$\left\{\frac{\hbar^2}{2\mu}\nabla^2_{R'} + E_{\text{rel}} - V(R')\right\}\chi(R') = \int K(R', R'')\,\chi(R'')\,dR''$$

$$R' = R_{\text{cl}} - R_{\text{tar}}. \tag{V,32}$$

_____

* It should be mentioned once more that the $\frac{3^+}{2}$ level of He[5] lies 0.11 MeV above the energy of a free deuteron plus a free triton.

This equation is valid for the region where the cluster and the target penetrate each other and also where they are separated, $K(\mathbf{R'}, \mathbf{R''}) = 0$. $E_{\text{rel}}$ is the relative kinetic energy of the cluster and the target. The target is now approximated by a nuclear matter wave function $\phi_{\text{n.m.}}$. Hence $E_{\text{rel}}$ is given by

$$E_{\text{rel}} = E_{\text{tot}} - (E_{\text{cl}}^{\text{bind}} + E_{\text{n.m.}}^{\text{bind}})    \qquad (V,33)$$

where $E_{\text{cl}}^{\text{bind}}$ and $E_{nm}^{\text{bind}}$ are the total binding energies of the cluster and the target if they are separated. The other notations in (V,32) are clear without further comment.

If one considers for instance the scattering of $\alpha$-clusters on nuclear matter and describes the $\alpha$-cluster by means of the usual oscillator wave functions and the nuclear matter wave function $\phi_{\text{m.n.}}$ by a Fermi sea wave function, then $V(\mathbf{R'})$ and $K(\mathbf{R'}, \mathbf{R''})$ can be calculated explicitly*. In this case $V(\mathbf{R'})$ and $K(\mathbf{R'}, \mathbf{R''})$ become translation invariant, i.e.

$$V(\mathbf{R'} + \Delta\mathbf{R}) = V(\mathbf{R'})    \qquad (V,34)$$

and

$$K(\mathbf{R'} + \Delta\mathbf{R},\; \mathbf{R''} + \Delta\mathbf{R}) = K(\mathbf{R'}, \mathbf{R''}) \,.$$

We assume further that the nuclear matter wave function, as well as the $\alpha$-cluster wave function, is rotationally invariant. It follows then that $V(\mathbf{R'})$ becomes a constant $V_0$, and $K(\mathbf{R'}, \mathbf{R''})$ becomes a function $K(|\mathbf{R'} - \mathbf{R''}|)$.

To obtain the optical potential for the scattering of $\alpha$-clusters on nuclear matter we assume that when the cluster is in the target then $\chi(\mathbf{R'})$ may be described by a plane wave "ansatz" of the form:

$$\chi(\mathbf{R'}) = e^{i\mathbf{k}\cdot\mathbf{R'}}    \qquad (V,35)$$

where $\mathbf{k}$ is the relative wave number vector when the cluster is inside the target nucleus.

Due to the invariance character of $V(\mathbf{R'})$ and $K(\mathbf{R'}, \mathbf{R''})$ the "ansatz" (V,35) must satisfy eq. (V,32). From the definition (V,33) of $E_{\text{rel}}$ it follows that $E_{\text{rel}}$ is a function of $\mathbf{k}^2$. The optical potential for the $\alpha$-nuclear matter scattering follows now from comparison of the following two equations, (V,36) and (V,37)

$$\left\{ \frac{\hbar^2}{2\mu} \nabla_{\mathbf{R'}}^2 + E_{\text{rel}}(k^2) \right\} e^{i\mathbf{k}\cdot\mathbf{R'}} = V_0 e^{i\mathbf{k}\cdot\mathbf{R'}} + \int K(|\mathbf{R'} - \mathbf{R''}|)\, e^{i\mathbf{k}\cdot\mathbf{R''}}\, d\mathbf{R''}$$

or

$$\left\{ -\frac{\hbar^2 k^2}{2\mu} + E_{\text{rel}}(k^2) \right\} = V_0 + \int K(|\mathbf{R'} - \mathbf{R''}|)\, e^{-i\mathbf{k}\cdot(\mathbf{R'} - \mathbf{R''})}\, d(\mathbf{R'} - \mathbf{R''})    \qquad (V,36)$$

and

$$\left\{ \frac{\hbar^2}{2\mu} \nabla_{\mathbf{R'}}^2 + E_{\text{rel}}(k^2) \right\} e^{i\mathbf{k}\cdot\mathbf{R'}} = V_{\text{opt}}\, e^{i\mathbf{k}\mathbf{R'}}$$

---

* Some further assumptions are made which have no essential influence on the result, see [83].

or

$$\left\{ -\frac{\hbar^2 k^2}{2\,\mu} + E_{\text{rel}}\,(k^2) \right\} = V_{\text{opt}}\,(k^2) \,. \tag{V,37}$$

Hence we have

$$V_{\text{opt}}\,(k^2) = V_0 + \int K(|\boldsymbol{R'} - \boldsymbol{R''}|)\; e^{-i\boldsymbol{k}\cdot(\boldsymbol{R'}-\boldsymbol{R''})}\,\mathrm{d}(\boldsymbol{R'} - \boldsymbol{R''}) \,. \tag{V,38}$$

As can be easily seen one obtains the same $V_{\text{opt}}$ if one describes $\chi(\boldsymbol{R'})$ as a linear superposition of plane waves with the same absolute value of the relative wave number vector.

Experimentally one measures $V_{\text{opt}}$ as a function of $E_{\text{rel}}$; therefore one has to express $k^2$, the argument of $V_{\text{opt}}$ above, as a function of $E_{\text{rel}}$. Fig. 19 shows $V_{\text{opt}}$ as a function of $E_{\text{rel}}$ for two-nucleon interaction forces of the form (V,17 and 17a).

The most remarkable feature of Fig. 19 is that for relatively low $E_{\text{rel}}$ values, $V_{\text{opt}}$ gets deeper as $E_{\text{rel}}$ increases. This looks strange at first because $V_{\text{opt}}$ is mainly determined by the total potential energy, which is the sum of the two-nucleon potential energies; and due especially to the Majorana force we expect the two-nucleon potential energy to become less deep with increasing scattering energy $E_{\text{rel}}$. However, this behaviour of $V_{\text{opt}}$ can be understood to arise from the Pauli principle as follows. Completely analogous to our considerations in chap. IV section D, the low momentum components of an $\alpha$-cluster wave function will be cut off by antisymmetrization if the cluster is in nuclear matter.

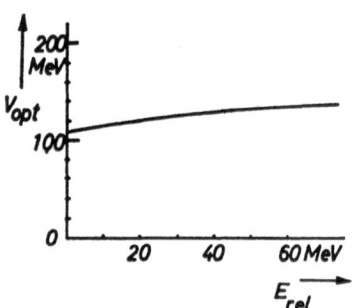

Fig. 19. Optical potential depth vs. energy for α-particle nuclear matter scattering

This cut-off effect causes the $\alpha$-cluster wave function (as in chap. IV sect. D) to spread out over a larger region, thereby lessening the internal potential energy of the $\alpha$-cluster compared to that of a free $\alpha$-cluster. With increasing $E_{\text{rel}}$ this cut-off effect is reduced and the internal potential energy of the $\alpha$-cluster increases. Hence at low scattering energy the total potential energy, and with it $V_{\text{opt}}$, actually becomes greater (i.e. $V_{\text{opt}}$ gets deeper) with increasing $E_{\text{rel}}$. For very high $E_{\text{rel}}$ values ($E_{\text{rel}} > 100$ MeV) this cut-off effect will gradually vanish and therefore $V_{\text{opt}}$ will become less deep again due to the decreasing two-nucleon potential.

$\alpha$-particle scattering experiments especially on $S^{28}$ indicate that this effect is real because these experiments seem to show that the phenomenological optical potential depth to fit these experiments becomes greater with the scattering energy if $E_{\text{rel}}$ is small ($E_{\text{rel}} < 10$ MeV). A comparison of the calculated and measured optical potential depths for higher scattering energies cannot be made because then the inelastic scattering (transitions to other reaction channels) becomes so large that the probability that the $\alpha$-cluster penetrates undisturbed into the target nucleus becomes very small* **.

Finally we want to mention that the depth of the central part of the experimentally determined optical potential agrees rather well with the calculated optical potential depth of nuclear matter. The calculated optical potential depth is about 10% smaller than the experimentally determined optical potential depth, R. DAVIS [170].

### b) Complex optical potentials

Until now we have considered optical potentials belonging to purely elastic scattering processes or to elastic scattering processes in which transitions to other reaction channels are neglected. As we mentioned at the beginning of this section, in the phenomenological theory of nuclear scattering and nuclear reaction processes it is very often useful to introduce complex optical potentials whose imaginary parts are essentially responsible for the description of transitions to other reaction channels (absorption processes). We will now very shortly discuss how such optical potentials can be derived from our coupled channel equations. For this we start by considering two coupled channel equations of the form (V,10). For simplicity we neglect as before spin-orbit coupling effects. This neglection again does not influence in any way the general conclusions which we will obtain.

If we abbreviate the operator $-\dfrac{\hbar^2}{2\,\mu_{\mathrm{I}}}\,\nabla^2_{\boldsymbol{R}'_{\mathrm{I}}} + V(\boldsymbol{R}'_{\mathrm{I}}) + \int\limits_{\boldsymbol{R}''} K(\boldsymbol{R}'_{\mathrm{I}}, \boldsymbol{R}''_{\mathrm{I}})$
with $H^0_{\mathrm{I}}$, the operator $\int\limits_{\boldsymbol{R}''_{\mathrm{II}}} K(\boldsymbol{R}'_{\mathrm{I}}, \boldsymbol{R}''_{\mathrm{II}})$ with $H^1_{\mathrm{I,II}}$, and so on, then we can write the equations (V,10) as the following operator equations

$$(H^0_{\mathrm{I}} - E_{\mathrm{I}})\,\chi(\boldsymbol{R}'_{\mathrm{I}}) = -H^1_{\mathrm{I,II}}\,\chi(\boldsymbol{R}''_{\mathrm{II}})$$
$$(H^0_{\mathrm{II}} - E_{\mathrm{II}})\,\chi(\boldsymbol{R}'_{\mathrm{II}}) = -H^1_{\mathrm{II,I}}\,\chi(\boldsymbol{R}''_{\mathrm{I}}) \,. \qquad (\mathrm{V,10a})$$

We consider now a reaction process where we have an incoming and outgoing particle beam in channel I and outgoing reaction products in channel II. We want to construct the complex optical potential which describes the scattering and absorption (transition to channel II) of the particle beam in channel I. To do this we introduce the channel II Green's function $G(\boldsymbol{R}''_{\mathrm{II}}, \boldsymbol{R}'''_{\mathrm{II}})$ to the operator $(H^0_{\mathrm{II}} - E_{\mathrm{II}})$ which satisfies the following inhomogenous equation

$$(H^0_{\mathrm{II}} - E_{\mathrm{II}})\,G(\boldsymbol{R}''_{\mathrm{II}}, \boldsymbol{R}'''_{\mathrm{II}}) = \delta(\boldsymbol{R}''_{\mathrm{II}} - \boldsymbol{R}'''_{\mathrm{II}}) \,. \qquad (\mathrm{V,39})$$

$G(\boldsymbol{R}''_{\mathrm{II}}, \boldsymbol{R}'''_{\mathrm{II}})$ has to satisfy the additional boundary condition that it contains only outgoing waves, because in the process considered only

---

* It should be remembered that antisymmetrization effects are not considered as disturbances here.

** In the language of the optical potential this means that the imaginary part of this potential — we shall discuss this imaginary part in the next paragraph — will become so large that the $\alpha$-cluster will be already absorbed near the surface of the target nucleus.

outgoing reaction products are present in channel II. With the Green's function $G(\mathbf{R}''_{II}, \mathbf{R}'''_{II})$ we can write for $\chi(\mathbf{R}''_{II})$:

$$\chi(\mathbf{R}''_{II}) = -G(\mathbf{R}''_{II}, \mathbf{R}'''_{II})\, H^1_{I,\,II}\chi(\mathbf{R}''_{I}) \tag{V,40}$$

or explicitly written out

$$\chi(\mathbf{R}''_{II}) = -\int\int G(\mathbf{R}''_{II}, \mathbf{R}'''_{II})\, K(\mathbf{R}'''_{II}, \mathbf{R}''_{I})\, \chi(\mathbf{R}''_{I})\, \mathrm{d}\mathbf{R}'''_{II}\, \mathrm{d}\mathbf{R}'' \tag{V,40a}$$

as can be verified easily by putting (V,40a) in the second equation of (V,10a).

If we put (V,40a) in the first equation of (V,10a), i.e. (V,10), then we obtain explicitly written out

$$\left\{-\frac{\hbar^2}{2\,\mu_I}\,\nabla^2_{\mathbf{R}'_I} + V(\mathbf{R}'_I)\right\}\chi(\mathbf{R}'_I) + \int\overline{K}(\mathbf{R}'_I, \mathbf{R}''_I)\,\chi(\mathbf{R}''_I)\,\mathrm{d}\mathbf{R}''_I = E_I\chi(\mathbf{R}'_I) \tag{V,41}$$

where here $\overline{K}$ is defined to mean

$$\overline{K}(\mathbf{R}'_I, \mathbf{R}''_I) \equiv K(\mathbf{R}'_I, \mathbf{R}''_I) - \int\int K(\mathbf{R}'_I, \mathbf{R}''_{II})\, G(\mathbf{R}''_{II}, \mathbf{R}'''_{II})$$
$$K(\mathbf{R}'''_{II}, \mathbf{R}''_I)\,\mathrm{d}\mathbf{R}'''_{II}\,\mathrm{d}\mathbf{R}''_{II}\,. \tag{V,41a}$$

Eq. (V,41) looks much like a usual scattering equation with a nonlocal interaction for the elastic scattering of the particles in channel I, as for instance eq. (V,6), only that the kernel which describes this nonlocal interaction looks now rather complicated. But a significant difference exists from scattering equations of the form (V,6) to those of (V,41). Unlike $K(\mathbf{R}', \mathbf{R}'')$ in (V,6), the kernel $\overline{K}(\mathbf{R}'_I, \mathbf{R}''_I)$ in (V,41) is no longer hermitian, that is, $\overline{K}(\mathbf{R}'_I, \mathbf{R}''_I)$ is *not* equal $\overline{K}^*(\mathbf{R}''_I, \mathbf{R}'_I)$. The reason for this is that the Green's function $G(\mathbf{R}''_{II}, \mathbf{R}'''_{II})$ contains only outgoing waves. This is responsible for the transitions from channel I to channel II, which appear in the incoming channel I as a partial absorption of the particles in channel I.

The optical potential belonging to (V,41) can now be constructed in the same way as the potentials for purely elastic scattering processes without transitions to other channels by comparing (V,41) with

$$\left\{-\frac{\hbar^2}{2\,\mu_I}\,\nabla^2_{\mathbf{R}'_I} + V_{\mathrm{opt}}(\mathbf{R}'_I)\right\}\chi(\mathbf{R}'_I) \equiv E_I\chi(\mathbf{R}'_I)\,. \tag{V,42}$$

As before, it is also possible here to split (V,41) and (V,42) into equations to given orbital angular momentum. Then $V_{\mathrm{opt}}(\mathbf{R}'_I)$ becomes, in addition to its energy dependence, $l$-dependent again.

This time the energy- and $l$-dependent $V_{\mathrm{opt}}$ will be complex with positive imaginary part due to the non-hermitian character of $\overline{K}(\mathbf{R}'_I, \mathbf{R}''_I)$; otherwise $V_{\mathrm{opt}}$ would not describe a partial absorption of the particles in channel I. Also the singularities of $V_{\mathrm{opt}}$ in its surface region will vanish due to the fact that now $V_{\mathrm{opt}}$ contains an imaginary part. But as long as the imaginary part of $V_{\mathrm{opt}}$ is not too large it will remain that $V_{\mathrm{opt}}$ varies very strongly in its surface region.

It is clear that the procedure which has been used here to obtain the optical potential for the incoming particles in the coupled two-channel case can be applied completely analogously to the coupled many-channel case. Here one has to use the Green's functions for each of the reaction channels to eliminate successively the relative motion wave-functions of the different reaction channels until one is left with a linear integro-differential equation of the form (V,41) which describes the scattering and partial absorption of the particles in the incoming channel. The resulting kernel $\bar{K}$ will again be non-hermitian and will become increasingly complicated to calculate with increasing number of channels considered. This linear integro-differential equation may be used in the same way as above to define the complex optical potential for the scattering and absorption of the particles in the incoming channel.

As we have already mentioned above, the many-channel calculations which one has to perform to calculate such optical potentials explicitly, starting from the two-nucleon interaction forces, become rather com-plicated even in the simplest cases. But some general features of such optical potentials for the particles in the incoming channel can already be seen in the one dimensional example mentioned at the end of section C above where the calculations can be carried out explicitly, [78]. It turns out that such complex optical potentials also vary relatively slowly with changing scattering energy. Only in energy regions where a sharp resonance level exists in one of the reaction channels do the real and imaginary part of the optical potential in the scattering channel change rather rapidly with changing scattering energy, see also [81]. Further it is clear that the influence of a special reaction channel on the real and imaginary part of the optical potential of the incoming channel will usually decrease with increasing number of open reaction channels.

Let us now make some summary remarks. We began with the $A$-nucleon Schrödinger equation. The nuclear wave function solving this equation was expanded in principle in an appropriately chosen complete (usually non-orthogonal) set of mixed cluster functions. To come to a practical approximate solution of the Schrödinger equation we retained only those few terms in the cluster expansion thought to be important on physical grounds. This limited "ansatz" was used as a trial function in the Ritz variational principle. The relative motion functions of the cluster functions were freely varied to come to the general set of coupled integro-differential equations for the relative motion functions (V,15), which we call the coupled channel equations. Solution of these equations then yields our approximate solution of the Schrödinger equation. Since a cluster expansion system is complete, we always have a definite, if lengthy, program whereby in principle we can improve our approximate solution step by step by including successive terms from the cluster expansion in our initial "ansatz". At the moment, the size of present day computers limits our "ansatz" to one or two cluster functions in practical calculations. Nevertheless we saw in section C with specific examples that we could obtain good quantitative results even with such simple trial functions. This is another justification of the importance of cluster correlations in nuclei.

Beyond practical calculations the general coupled channel cluster formulation leads to important qualitative conclusions concerning the behaviour of nuclei. For instance from certain examples treated here we have seen that direct reaction theory is contained within our general formulation. The principle difference between the (completely antisymmetrized) reaction theory outlined here and such direct reaction theories using partially distinguishable nucleons (incomplete antisymmetrization) is that the purely local interaction between the composite particles in such theories is replaced here by a non-local interaction (integral kernels) between the clusters. The non-local interaction arises primarily from the antisymmetrization between the clusters.

We saw also that our general coupled channel equations provide a theoretical basis for the phenomenological optical model and distorted wave description of nuclear reactions. In section $D$ we saw how to derive a (usually complex) optical potential in the incoming channel to replace the effects of the different local and non-local interactions between the clusters. In a similar way one can also derive optical potentials for the various reaction channels to describe the distortion of the outgoing wave function for the reaction products. In specific examples we saw that, with the exception of energies near resonances in other channels, these optical potentials varied quite slowly with bombarding energy of the incident cluster, [78, 81]. Hence the phenomenological optical model formalism is contained in our formulation.

We see further that the optical potentials of distorted wave calculations are not uniquely defined. This is connected with antisymmetrization. The optical potentials are derived from the local and non-local interactions $V(\mathbf{R}')$ and $\overline{K}(\mathbf{R}', \mathbf{R}'')$ of (V,41). The potential $V(\mathbf{R}')$ contains the direct terms and $\overline{K}(\mathbf{R}', \mathbf{R}'')$ all exchange terms. However what we call direct and exchange terms depends on how we describe the target nucleus. For instance in the $F^{19}$ (p, $\alpha$) $O^{16}$ reaction, making the oscillator assumption, it depends on whether we represent $F^{19}$ with an $O^{16}$-$t$ or $N^{15}$-$\alpha$ cluster structure. In general the resulting optical potential will be different for such different cluster representations of the same target. However the distorted wave solutions for these different optical potentials will all be identical in the asymptotic region.

A further application is that we can use the connection between the general cluster formulation and these phenomenological theories sketched here to shed light upon the complicated behaviour of the coupled channel equations themselves. For instance the optical model description of the differential elastic scattering cross-section for $\alpha$-He$^3$ elastic scattering at 16.6 MeV can be greatly improved by adding to the real potential an imaginary part to account for transitions to other channels which occur at this energy. This means to the hermitian kernel of the integrodifferential equation describing pure $\alpha$-He$^3$ elastic scattering (see sect. C subsect. b) we should add a non-hermitian term phenomenologically chosen so that together with the hermitian part we obtain the correct inelastic $\alpha$-He$^3$ cross-section. In principle this non-hermitian term could be deduced by including the inelastic channels in our initial cluster "ansatz" and solving the resulting coupled channel equations.

With this we conclude the discussion formulating nuclear reaction theory in terms of coupled channel equations by the use of cluster representations. In this formulation the description of nuclear bound states and the description of nuclear reactions differ only in the boundary conditions applied to the coupled channel equations. Hence as we promised earlier, the method of cluster representations provides a unified approach connecting nuclear reaction theory and nuclear structure theory in a very natural way. By starting from the beginning with completely antisymmetrized wave functions (including the incident particle), using mixed cluster representations and not introducing boundary conditions at the surface of the target nucleus one avoids many difficulties which appear in other reaction theories, see for instance [62], G. BREIT and E. WIGNER [84], P. KAPUR and R. PEIERLS [85], H. FESHBACH [86], J. HUMBLET and L. ROSENFELD [87], W. MACDONALD [88].

# VI. Introduction of Hard-Core Forces

## A. General remarks

The unified theoretical description of nuclear bound states and nuclear reactions discussed up to this point is in one respect very unsatisfactory. The saturation character of the nuclear forces has always been described in a very crude way by describing the internal structure of the clusters with internal functions having unvaried internal width parameters*. We will now discuss how the above description of nuclear bound states and nuclear reactions can be refined to include the saturation character in the calculations in a theoretically consistent manner.

To adequately fit the two-nucleon scattering data above 200 MeV requires a more sophisticated phenomenological potential than any we have used so far in our calculations. An adequate fit is obtained by the inclusion of either a hard-core or a velocity-dependent force in the two-body potential. If we are not to violate our original approach that we use forces describing the two-nucleon data satisfactorily, then we must include such forces in our phenomenological potential. We shall show here that the inclusion of hard-core forces are sufficient to give saturation**.

If we include hard-core forces in our nuclear Hamiltonian (see eq. (VI,2) for an explicit form) then we must modify our trial wave functions so that no two nucleons may penetrate their mutual hard core, K. WILDERMUTH [90]. Formally this can be done very simply by multiplying the cluster wave functions with a correlation (Jastrow) factor of the

---

    * With this one avoids the collapse of the clusters.
    ** Saturation has also been obtained with velocity-dependent forces; for a comparison of the two see E. SCHMID et al. [89].

form [91]

$$\phi_{corr} = \Pi_{\text{all pairs}}\, f\left(|r_i - r_k|\right) \tag{VI,1}$$

with the property that for values of $|r_i - r_k|$ smaller than the range of the hard core the factor $f(|r_i - r_k|)$ is zero. When $|r_i - r_k|$ becomes large this factor must go to unity. In Fig. 20 the qualitative behaviour of such a factor is given as a function of $r = |r_i - r_k|$.

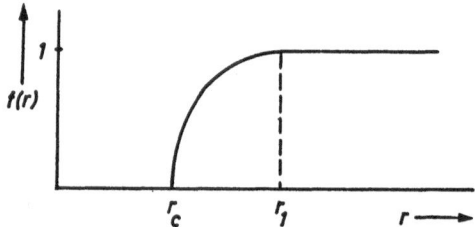

Fig. 20. Behaviour of cut-off function $f(r)$ vs. nuclear separation distance

In Fig. 20 it was assumed that the hard-core range of the nuclear forces is $0.4 \cdot 10^{-13}$ cm. With correlation factors of the form (VI,1) in the cluster functions all integrals which appear in the expectation value of the Hamiltonian converge even if one uses realistic two-nucleon interactions with hard-core forces.

This correlation factor is rotationally and translationally invariant. Therefore its introduction does not disturb the geometrical properties of the generalized cluster functions. That means if the considered cluster function is an eigenfunction to the total spin, isobaric spin, parity, momentum, and so on, before the introduction of such a correlation factor it will remain so after the introduction of such a factor. If $f(|r_i - r_k|)$ is the same function for all nucleon pairs, then $\phi_{corr}$ of (VI,1) is symmetric in all nucleon coordinates and has not to be included in the antisymmetrization. A further consequence is that equivalences between different cluster functions remain valid even after the introduction of such a correlation factor. For example in the oscillator model the ground state Li⁶ description as *a-deuteron* clusters or as *triton-He³* clusters will not be destroyed by the factor.

The calculations can usually be refined further by choosing different functions $f_{ik}(|r_i - r_k|)$ for different nucleon pairs. In this case $\phi_{corr}$ is no longer symmetric in all nucleon coordinates and therefore one has to include $\phi_{corr}$ in the antisymmetrization, but nothing else need be changed in principle.

From these remarks we see that the introduction of correlation factors of the form (VI,1) in the cluster functions and of hard-core forces in the Hamiltonian does not change in principle the considerations we developed in the first five chapters. In particular the unified general formulation of nuclear bound states and nuclear reaction processes remains completely valid. Certainly the introduction of such factors complicates the calculations, especially of the integral kernels. One important reason for this is that $f(|r_i - r_k|)$ is a non-analytic function (since it vanishes identically in a finite region). An important point for the calculations with hard-core forces is that one has to include in the correlation factors of the form (VI,1) variational parameters which determine especially the slope of $f(|r_i - r_k|)$ near the hard-core distance, because the expectation value of the energy usually depends very sensitively on this slope [89].

## B. The ground state of Li⁶ and its charge distribution

As already pointed out earlier Li⁶ can be considered as a nucleus on which one can test improvements in the theoretical description of nuclear phenomena. Therefore we shall discuss some calculations and their results concerning the ground state of Li⁶ and its reactions in which hard-core effects are now explicitly included, E. Schmid et al. [105].

In this respect the charge distribution of Li⁶ is of special interest. Analysis of the experimental results shows that Li⁶ has a large r.m.s. radius and a very diffuse surface as compared to other $p$-shell nuclei, see for instance U. Meyer-Berkhout et al. [92]. Its charge distribution has a long tail which cannot be described, for instance, by using single particle wave functions in an infinite oscillator potential, G. Burleson and R. Hofstadter [93], D. Jackson [94], L. Elton [95]. This is in direct contrast to the fact that for all other $p$-shell nuclei the experimental results can be fitted fairly well using wave functions in such a potential well [92].

The diffuseness of Li⁶ seems to be a typical clustering effect. For the ground state of Li⁶ we have from our earlier considerations that the predominant cluster structure is the $\alpha$-*deuteron* structure, with other structures such as $He^3$-triton playing only a minor role*. In the $\alpha$-*deuteron* structure the two clusters are bound together by only 1.45 MeV, which means that they are on the average rather far apart and behave more or less like free particles. We can thus expect the *deuteron* cluster will have a long tail, just as a free deuteron does. Qualitatively we can therefore understand the diffuseness and large r.m.s. radius of Li⁶ as arising from the long tail of the *deuteron* cluster and the fact that the clusters are relatively widely separated.

To confirm these qualitative arguments and to show that the generalized cluster description of Li⁶ is flexible enough to describe these features** a variational calculation was carried out with the following two-body

---

* This is also confirmed by experimental evidence, [37]. Further references on the $\alpha$-deuteron structure of Li⁶ besides [29], [31], [32] are J. Dabrowski and J. Savicki [96], I. Vashakidze and G. Chilashvili [97], T. Kapaleishvili [98].

** It is in just such cases that the flexibility of the cluster description to incorporate our physical knowledge shows its power. In an arbitrary complete set the same effect could be achieved in principle by mixing in the proper amount of many higher excited states; but it would be virtually impossible in practice to determine the correct admixture, since the large number of variational amplitudes required would be without direct physical meaning.

hard-core potential, where $r_{ik} = |r_i - r_k|$ is the interparticle distance:

$$V_t(r_{ik}) = \begin{cases} +\infty & ; \ r < r_c \\ -V_t e^{-\alpha(r_{ik}-r_c)}; & r \geq r_c \end{cases}$$

$$V_s(r_{ik}) = \begin{cases} +\infty & ; \ r < r_c \\ -V_s e^{-\beta(r_{ik}-r_c)}; & r \geq r_c \end{cases}$$

$$V_t = 434.0 \text{ MeV} \qquad\qquad V_s = 216.0 \text{ MeV}$$

$$\alpha = 2.4 \cdot 10^{13} \text{ cm}^{-1} \qquad\qquad \beta = 1.97 \cdot 10^{13} \text{ cm}^{-1} \qquad (\text{VI},2)$$

$$r_c = 0.35 \cdot 10^{-13} \text{ cm} .$$

In the above equation $V_t(r)$ and $V_s(r)$ are the triplet and singlet two-nucleon interactions in the even orbital angular momentum states. The interaction in the odd orbital angular momentum states is taken to be zero except for a hard core of radius $r_c$. This potential describes the two-nucleon scattering data up to several hundred MeV rather well [89]*.

In order to obtain a reliable r.m.s. radius from the variational calculation, a rather flexible trial function was assumed, having the following form

$$\psi(Li^6) = \mathscr{A} \, \phi_\alpha(\bar{r}_1 \dots \bar{r}_4) \, \phi_d(\bar{r}_5 \bar{r}_6) \, \chi(R) \, \xi(s_1 \dots t_6) \prod_{\substack{n=1,2,3,4 \\ m=5,6}} f_2(r_{nm}) \qquad (\text{VI},3)$$

where the operator $\mathscr{A}$ stands again for total antisymmetrization and $\xi(s_1 \dots t_6)$ is a spin charge function chosen to give $S = 1$ and $T = 0$ for the wave function $\psi(Li_6)$.

The $\alpha$-cluster internal function is taken as, [105],

$$\phi_\alpha = \prod_{i>k=1}^{4} f_1(r_{ik}) \left\{ e^{-\frac{a_1}{8} \sum_{i>k=1}^{4} r_{ik}^2} + a_2 e^{-\frac{a_2}{8} \sum_{i>k=1}^{4} r_{ik}^2} \right\}. \qquad (\text{VI},4)$$

The deuteron function $\phi_d$ is generated by the Schrödinger equation

$$-\frac{\hbar^2}{M} \frac{d^2}{dr^2} (r\phi_d(r)) + (V_t(r) - b_2) \, r \phi_d(r) = 0 \qquad (\text{VI},5)$$

between $r = 0$ and $r = b_1$. For $r > b_1$, the function

$$\phi_d(r) = \frac{B_1}{r} [e^{-b_3 r} - B_2 e^{-B_3 r}] \qquad (\text{VI},5a)$$

is used. The constants $B_1$, $B_2$ and $B_3$ are determined by the condition that $\phi_d$ and its first two derivations are continuous at $r = b_1$. With this trial wave function for the deuteron cluster the tail can be varied without any variation in the interior region of the deuteron cluster. This seems to be a desirable feature since the interaction of the clusters might conceivably alter the tail of the deuteron cluster function without having much

---

* We neglect again spin-orbit forces, which due to the spherical symmetry of the assumed variational wave function for Li⁶ would only indirectly (by changing the constants of the central forces somewhat) influence the results a little.

effect in the interior region of the cluster where the neutron-proton interaction dominates.

The long range part of the relative motion function is chosen as

$$\chi(\boldsymbol{R}) = R^2 (e^{-c_1 R^2} + c_2 e^{-c_3 R^2})\, Y_{00}(\theta_R,\, \varphi_R) \qquad (VI,6)$$

where $\boldsymbol{R} = \boldsymbol{R}_\alpha - \boldsymbol{R}_d$ is the vectorial separation distance of the two clusters. The inclusion of a second Gaussian function is necessary, since the binding between the two clusters is rather weak. Therefore a single Gaussian function will not be able to represent the behaviour of the relative motion properly.

The cut-off functions $f_1$ and $f_2$ are equal to zero within the hard-core region. Outside they are generated by the differential equation

$$-\frac{\hbar^2}{M}\frac{d^2}{dr^2}(r f_{1,2}(r)) + \frac{1}{2}\gamma_{1,2}(V_t(r) + V_s(r)) r f_{1,2}(r) = \varepsilon_{1,2} r f_{1,2}(r) \quad (VI,7)$$

up to the value $r = r_{1,2}$ where $f_{1,2}$ has its first maximum; the maximum is normalized to unity. For values of $r$ larger than $r_{1,2}$ the function $f_{1,2}$ is taken to be equal to one. The variational parameters $\gamma_{1,2}$ are essentially responsible for the cut-off factors near the hard-core distances, i.e. the initial slope at $r = r_c$.

The expectation value of the six-body Hamiltonian

$$\overline{H} = \overline{T} + \frac{1}{2}\sum_{i \neq k}^{6} V_{ik} \qquad (VI,8)$$

was calculated by a Monte Carlo method*, E. SCHMID [99].

A minimum was found with the following parameters:

$$a_1 = 1.17 \cdot 10^{26}\ \text{cm}^{-2};\ a_2 = 0.25;\ a_3 = 0.52 \cdot 10^{26}\ \text{cm}^{-2}$$

$$b_1 = 1.35 \cdot 10^{-13}\ \text{cm};\ b_2 = -2.22\ \text{MeV};\ b_3 = 0.34 \cdot 10^{13}\ \text{cm}^{-1}$$

$$c_1 = 0.18 \cdot 10^{26}\ \text{cm}^{-2};\ c_2 = 0.25;\ c_3 = 0.065 \cdot 10^{26}\ \text{cm}^{-2}$$

$$\gamma_1 = 0.68;\ \gamma_2 = 1.0;\ \varepsilon_1 = 0;\ \varepsilon_2 = 0\,. \qquad (VI,9)$$

We see that this wave function reflects the saturation character of the nuclear forces quite satisfactorily, and we emphasize that no parameters of this wave function were fixed phenomenologically; all were freely determined by the variation**. Hence the saturation is now the result of a completely consistent theoretical treatment beginning with the six-body Schrödinger equation including hard-core forces.

---

\* About 50 hours of computing time was needed on the IBM-7090 computer.

\** We emphasize that we no longer use the trick of fixing the internal width parameters of the clusters phenomenologically to account for saturation. Here the internal width parameters along with the other parameters are varied completely freely to find the minimum of $\langle H \rangle$. Thus we have taken into account both the change of the cluster "radius" and the anharmonicity effect mentioned in chapter IV section A in a theoretically consistent way.

The parameters for the $\alpha$-cluster wave function are the same as those for a free $\alpha$-particle [89]. This indicates that the $\alpha$-cluster has a very small compressibility as is to be expected. The *deuteron* cluster has a somewhat shorter tail than a free deuteron*. The relative motion function is rather long ranged.

The energy at the minimum is $-31.3 \pm 0.8$ MeV (experimentally $= -32.0$ MeV), of which 0.5 MeV (experimentally $-1.43$ MeV) represents the interaction energy between the clusters (separation energy of Li$^6$ into an $\alpha$-particle and deuteron). Although the interaction energy is rather small, the minimum is well defined because of the Coulomb barrier which is about 1 MeV high. The r.m.s. radius of the charge distribution is $2.73 \pm 0.15 \cdot 10^{-3}$ cm, with the error coming from the statistical uncertainty in the position of the minimum. A finite charge distribution of the proton with a Gaussian shape and a r.m.s. radius of $0.72 \cdot 10^{-13}$ cm has been included in this value. The r.m.s. radius derived from the electron scattering experiments is $2.70 \pm 0.15 \cdot 10^{-13}$ cm [92]. Fig. 21 shows a comparison between the theoretical charge distribution and the phenomenological charge distributions derived from electron scattering experiments under various assumptions.

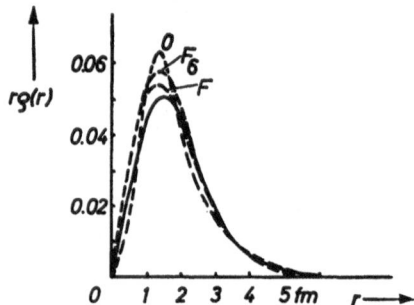

Fig. 21. Charge distributions of Li$^6$, including finite charge distribution of proton. Solid line: theoretical curve according to optimum wave function; dotted line (phenomenological curves): $O =$ oscillator model, $F =$ Fermi model, $F_6 =$ Fourier model with cut-off at $6 \cdot 10^{-13}$ cm, see [12]

As is seen, the agreement is quite good in the region of the tail which contributes most to the r.m.s. radius. In this region only the curve labeled "$F_6$" shows a relatively large deviation from the calculated one, but this is because this curve was derived under the assumption of zero charge density beyond $6 \cdot 10^{-13}$ cm. In the interior region the experiments do not give accurate information. The analysis of [92] can only conclude that the maximum charge density in Li$^6$ is $0.064 \pm 0.006 \cdot 10^{39}$ protons per cm$^3$.

In calculating expectation values and transition matrix elements with wave functions of the form (VI,3) most of the difficulties arise from its

---

* In the deuteron cluster wave function, $B_3$ is much larger than $b_3$ and $B_2$ is rather small; hence the asymptotic behaviour of this function is determined by $b_3$. For the deuteron cluster, the optimum value of $b_3$ is 0.34, larger than the corresponding value of 0.23 for a free deuteron.

cut-off part. Many one-particle operators, however, are insensitive to short-range many-particle correlations, and therefore the cut-off part can be disregarded in calculating their expectation values and transition matrix elements. As an example the proton distribution of Li⁶ with and without cut-off was calculated. The result is shown in Fig. 22.

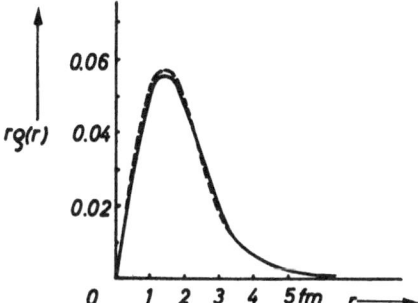

Fig. 22. Influence of hard core on charge distribution; solid line: according to optimum wave function, dotted line: same without hard-core correlation factor

As is seen the cut-off functions have only minor influence on the expectation value of the proton density operator.

## C. The Coulomb-disintegration of nuclei

As an example of nuclear reactions in which we use nuclear wave functions calculated with hard-core forces, we discuss next the Coulomb-disintegration of Li⁶ into an $\alpha$-particle and deuteron, J. HANSTEEN and H. WITTERN [100]. This will provide still another check on the quality of this refined cluster wave function.

The disintegration of Li⁶ into $\alpha$-particles and deuterons was observed by the bombardment of a gold target with Li⁶ nuclei, R. OLLERHEAD et al. [101] and C. ANDERSON et al. [102]. The bombarding energies were in the region of 25 to 60 MeV. Because these bombarding energies were largely below the Li⁶-gold Coulomb barrier it was concluded that what had been observed was due to the Coulomb-disintegration of Li⁶. This conclusion has a very interesting consequence because in the theoretical calculation of such a process one need therefore take into consideration only the inital and final state of the reaction products and the well-known Coulomb interaction. This means the theoretical description of Li⁶ Coulomb-disintegration and its comparison with experiment can give us a direct insight into the detailed structure of the Li⁶ ground state and some of its excited states.

In the first theoretical calculation of the total cross-section of this process Li⁶ was described in the frame of the oscillator shell model assuming a reasonable width parameter for the Li⁶ oscillator shell model wave function, with the result that the theoretical cross-section was by a

factor 10 smaller than the experimental cross-section, J. HANSTEEN and
I. KANESTRÖM [103]. This indicates again that the correct $Li^6$ description
deviates appreciably from the oscillator shell model description. We will
now show that we obtain the total $Li^6$ Coulomb-disintegration cross-
section within the experimental errors if we use for the ground state of
$Li^6$ the above described refined cluster wave function and for its excited
states cluster wave functions which can be derived from the $Li^6$ ground
state wave function.

The semi-classical treatment of the Coulomb-excitation of nuclei [56]
is readily adopted to this disintegration process. This means we describe
the path of $Li^6$ in the Coulomb field of the gold target by its classical
orbit and neglect the effect of the $Li^6$ excitation or disintegration on this
path. These approximations are justified if the following two conditions
are fulfilled, N. BOHR [104]:

First we must have

$$\frac{2 Z_1 Z_2 e^2}{\hbar v} \gg 1 \qquad\qquad (VI,10)$$

where $Z_1$ and $Z_2$ are the charge numbers of the projectile and the target
respectively, and $v$ the velocity of the projectile in the laboratory system.

Secondly it must be that

$$\frac{\Delta E}{E} \ll 1 \qquad\qquad (VI,11)$$

where $\Delta E$ is the c. m. energy loss suffered by the projectile in the
excitation or break-up process, $E$ being the bombarding energy.

The second condition is clear without further comment. The first
condition is equivalent to demanding, as can be shown easily, that along
the entire path of the $Li^6$ nucleus $r(t)$ in the Coulomb field of the gold
target we have

$$\frac{\mathrm{d} V}{\mathrm{d} r} \lambda_{Li^6} \ll V \qquad\qquad (VI,12)$$

fulfilled. Here $V$ is the $Li^6$-gold Coulomb potential and $\lambda_{Li^6}$ is the local
W.K.B. wavelength of $Li^6$. This means the change of the above Coulomb
potential is relatively very small across one $Li^6$ wavelength. An estimation
shows that for the $Li^6$ bombarding energy range considered here both
conditions are fulfilled in good approximation*. The time-dependent
interaction energy which is responsible for the Coulomb-excitation has
therefore the following form

$$H_{\mathrm{int}} = Z_2 e^2 \sum_{j=1}^{6} \left( \frac{1}{|r_j - r(t)|} - \frac{1}{|r(t)|} \right) \frac{1}{2} (1 + \tau_z^j) \qquad (VI,13)$$

The coordinate system is chosen in such a way that $r_j$ represents the
distance vector of the $j^{\mathrm{th}}$ nucleon from the $Li^6$ center of mass point.
$r(t)$ represents the time dependent position of this center of mass point

---

* This estimation shows also that the approximate calculation of the total
Coulomb-disintegration cross-section differs from the exact one by less than half
of a percent. This is in contrast to the differential cross-sections, where the correc-
tions are of the order of several percent.

from the target nucleus. The operator $\tau_z^j$ is the third component of the one-particle isobaric spin operator, and $Z_2$ is the charge of the target nucleus. In equation (VI,13) the Coulomb interaction energy between $Li^6$ and the gold center of mass point is subtracted because this interaction energy is reponsible for the hyperbolic orbit of the $Li^6$-gold relative motion described by $r(t)$.

The calculation of the total cross-section of the $Li^6$ Coulomb-disintegration is done in the usual way by means of time-dependent perturbation theory. For this calculation one has to observe the following points:

First: $Li^6$ decays mainly by electric quadrupole excitation, because it is the lowest possible Coulomb-excitation mode. Dipole excitation cannot happen because in the initial and final states the center of charge coincides with the center of mass of $Li^6$.

Second: $Li^6$ can decay directly into a continuum state or it can be excited at first into one of the low excited $Li^6$ triplet states $I = 3, I = 2, I = 1$ with $T = 0$ (see Fig. 3) and then decay into a continuum state. An estimation shows that the $\gamma$-transition probability from these latter states back to the ground state can be neglected compared to the decay probability of $Li^6$ into an $\alpha$-particle and a deuteron.

The decay of $Li^6$ directly into a continuum state and through the just mentioned triplet states can be treated in very good approximation independently from each other. As already mentioned in chapter IV section E the triplet states of $Li^6$ can be considered also in good approximation as rotational states to the $Li^6$ ground state. That means we can use for the description of these states the wave function (VI,3) only changing in the relative motion and spin isospin functions

$$\chi(\boldsymbol{R})\,\xi = \chi(|\boldsymbol{R}|)\,Y_{lm}(\theta_R,\,\phi_R)\,\xi\,(s_1 \ldots t_6) \qquad (VI,14)$$

the orbital and spin angular momentum part in such a way that we obtain eigenfunctions to the total angular momentum $I = 3, 2, 1$ with the orbital angular momentum value of the relative motion $l = 2$. The radial part $\chi(|\boldsymbol{R}|)$ in (VI,14) remains the same as in (VI,6). The $\alpha$-deuteron relative motions in the continuum states of $Li^6$ can be described by pure Coulomb wave functions because with the small decay energies of $Li^6$ considered here the phase shifts in the $\alpha$-deuteron scattering states caused by the $\alpha$-deuteron nuclear interaction can be neglected in good approximation for scattering energies off the resonances. Further in the initial and final states of $Li^6$ we can disregard the cut-off part of the wave functions because the Coulomb interaction operator (VI,13) consists of a sum of long ranged one-particle operators. In the continuum $Li^6$ states we describe the free deuteron-cluster by a Hulthén wave function [72].

If one computes with the above assumptions (and some further physically unessential simplifications, see for this [100]) the total Coulomb-decay cross-section of $Li^6$, then one obtains the results shown in Fig. 23. One sees the theoretical results lie within the error range of the experimental results, i. e. the agreement betwenn theory and experiment is really good. The calculations show further that the main contribution to the total cross-section comes from the transition to the $3^+$ resonance state.

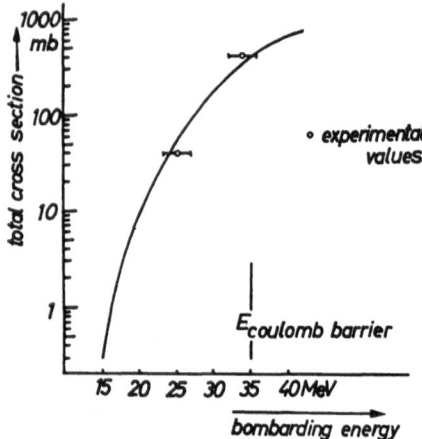

Fig. 23. Coulomb-disintegration cross-section vs. bombarding energy for Li⁶ on gold

We repeat that all calculations where $Li^6$ is described by means of shell model wave functions give a Coulomb-decay cross-section which is by a factor 100 too small. The main reason for this is that in the shell model description of $Li^6$ the amplitude of the *a-deuteron* relative motion is much too small. Due to the long ranged quadrupole interaction this reduces very much the calculated decay cross-section.

A further remark should be mentioned. In the calculations of the transition matrix elements with unnormalized $Li^6$ cluster wave functions the exchange terms contribute only very little ($<1\%$). They are consequently omitted. That this can be done arises from the smallness and the relatively slow quadratic increase of the quadrupole interaction for small *a-deuteron* cluster distances in the here considered $Li^6$-decay energy range. However, in the normalization terms the exchange terms cannot be omitted. Calculations show their contributions to be of the order of ten percent or less depending on the angular momentum of the state [100].

Finally we want to stress once more that no data fitting whatever has been made. However the simplifications we have made, i. e. dropping the cut-off factor and partial omission of the antisymmetrization, are only permissible under the very special aspects of the process under study. For instance one can also use the refined cluster wave function of the $Li^6$ ground state discussed here for the investigation of (p, 2p) scattering on $Li^6$. But in this case the neglection of the cut-off factors and the antisymmetrization in the transition matrix elements is much more doubtful because the nuclear interaction which is responsible for this reaction is very short ranged.

## VII. Qualitative Considerations

In the preceding chapters we have discussed how with the aid of the Ritz variational principle and using generalized cluster functions as trial functions one can come to a unified description of nuclear bound states and nuclear reactions. This method is very flexible because one can select the most suitable cluster function system for each problem which one wants to consider. On the other hand if one wants to describe a nuclear bound state or a nuclear reaction quantitatively, then the amount of calculation necessary increases factorially with the increasing number of nucleons and cluster structures which participate in the state or reaction considered. This is especially true if one uses realistic two-nucleon forces including a hard-core for the calculations as was done in chapter VI. Therefore up to the present these calculations have been carried out explicitly only for a restricted number of states and reactions in light nuclei. Nevertheless we can obtain valuable physical insight into the structure of nuclei and their reaction mechanisms by qualitatively applying the physical concepts which we have developed in the preceding chapters to problems where quantitative calculations are not practically possible at present, even with computers. The situation may be compared with that in molecular physics, where quantum mechanics is believed to give a correct description of the interactions between the electronic structures of the constituent atoms but quantitative calculations are very often impossibly difficult. Nevertheless great progress in understanding molecular structure has resulted from the more or less qualitative application of quantum mechanical ideas. In the remaining pages of this article we will therefore demonstrate on several examples the kind of physical insight which can be obtained by qualitative application of the previously developed concepts.

### A. Optical giant resonances in nuclear reactions

*a) Optical resonances in nucleus-nucleus scattering*

Let us start by stating some general qualities of the optical model for nucleon-nucleus scattering, H. FESHBACH et al. [106]. In the scattering of nucleons on nuclei the dependence of the averaged elastic and reaction cross-sections on bombarding energy and the nuclear radius can be approximately described by a phenomenologically chosen model potential. This so-called optical potential is complex and has the form

$$V = - (V_0 + i V_1) . \tag{VII,1}$$

Roughly, $V_0$ is responsible for the elastic scattering and $V_1$ for the inelastic scattering. For the occurrence of optical resonance phenomena (giant resonances) and optical interference phenomena in the elastic and reaction cross-sections it is necessary that the lifetime of the scattering nucleon within the (unexcited) nucleus be of the order of, or larger than, the time of traversal of the incident nucleon across the nucleus. This is equivalent to saying that the mean free path of the incident nucleon in the nucleus must be of the order of, or larger than, the size of the target nucleus. It is also important that the optical potential (VII,1) have a refractive edge, i. e. that the shape of the potential undergoes a rather abrupt change at the nuclear surface. When the optical potential does not change abruptly at the nuclear surface, the reflection probability there will be small and the momentum transfer from the nucleus to the scattering nucleon will not have distinct maxima characteristic of resonance phenomena, see for instance J. NODVIK [107]. For example optical resonance phenomena do not occur for a Gaussian potential.

We briefly summarize the general features of the resonance phenomena associated with nucleon-nucleus scattering as follows. The probability per unit time for exciting the nucleus, as can easily be derived, is approximately

$$W \approx \frac{2V_1}{\hbar} \qquad \qquad \text{(VII,2)}$$

giving for the partial energy width

$$\Delta E \approx 2 V_1. \qquad \qquad \text{(VII,3)}$$

For most nuclei $V_0$ is about 40–50 MeV and $V_1$ is about 3–5 MeV, [106]. For nuclei around $A = 200$ and $R \approx 8 \cdot 10^{-13}$ cm, these values correspond to an energy separation of about 50 MeV between giant resonances with the same angular momentum and a partial energy width of the resonances, from eq. (VII,3), of about 5–10 MeV. The corresponding lifetime $\Delta t = \hbar/\Delta E$ is about $10^{-22}$ sec, a value which is the same order of magnitude as the traversal time of the scattering nucleon across the nucleus. Since these times are comparable, the total energy width of the optical giant resonances is also about 5–10 MeV. The values of $V_0$ and $V_1$ for nucleon-nucleus scattering calculated from two-nucleon interactions similar to (III,14) and (V,17) are in approximate agreement with the phenomenological values given above, M. GOLDBERGER [108], A. LANE and C. WANDEL [109], P. MITTELSTAEDT [110], K. BRUECKNER et al. [111]. The calculations also indicate that $V_0$ decreases with increasing bombarding energy.

From our considerations in chapter V section D it is obvious that the general qualities of the optical model stated above are contained implicitly in the nuclear reaction theory sketched in chapter V.

We now ask whether it is reasonable within the framework of the optical model to expect resonance phenomena to occur in the scattering of composite particles such as $\alpha$-particles, tritons, etc., on nuclei, K. WILDERMUTH and R. CAROVILLANO [113]. This question is particularly interesting since certain experiments vividly display such optical phenomena, see for instance H. MELKANOFF [112].

From a "naive" theoretical point of view these experimental results are at first sight quite unexpected. For example consider the scattering of $\alpha$-particles on nuclei. As stated above, for a giant resonance to occur it is necessary that the lifetime of the $\alpha$-particle in the unexcited nucleus be of the order of, or larger than, the time of traversal across the nucleus, (see below for a more exact discussion of lifetime, etc.). When the $\alpha$-particle touches the nuclear surface, the average kinetic energy $E_{kin}$ of each constituent nucleon in the compound nuclear system is about 30 MeV. This follows directly from the Weizsäcker-Bethe formula together with the Fermi gas model for nuclei. Thus we can say that the time required for an exchange of a nucleon in the $\alpha$-particle with a nucleon in the nucleus is the order of magnitude of

$$\tau = \frac{R}{\sqrt{2E_{kin}/M}} \approx 2 \cdot 10^{-23} \text{ sec}. \tag{VII,4}$$

If the time given by eq. (VII,4) were a correct estimate of the lifetime of the $\alpha$-particle in the nucleus, as was often argued some years ago, it would be considerably smaller than the time of traversal of the $\alpha$-particle across the nucleus and would seem to preclude the possibility of any optical resonance effect. The inclusion of centrifugal barrier effects leaves this result essentially unchanged, as can be seen easily.

In the above discussion, however, the very important influence of the indistinguishability of the nucleons has not been considered. One cannot simply use (VII,4) to obtain an estimate of the lifetime of the scattered (composite) particle in the nucleus. Due to the considerations in the preceding chapters the lifetime of the possible nuclear states must be derived from the following kind of (cluster) wave function:

$$\psi = \mathscr{A}\{\phi_B \, \phi_T \, \chi \, (\boldsymbol{R}_B - \boldsymbol{R}_T)\} \tag{VII,5}$$

$\phi_B$ and $\phi_T$ describe the internal degrees of freedom of the bombarding and target clusters respectively, while $\chi \, (\boldsymbol{R}_B - \boldsymbol{R}_T)$ denotes the dependence on the relative separation of the centers of mass of these clusters. Due to the total antisymmetrization of (VII,5), indicated by the operator $\mathscr{A}$, the exchange effects considered in eq. (VII,4), which occur in about $10^{-23}$ sec, do not determine the lifetime of the nuclear clusters (i. e. the time before the cluster structure (VII,5) is destroyed). Each such exchange of nucleons merely introduces a minus sign before the wave function (VII,5) and therefore does not destroy the nuclear cluster wave function (VII,5)*. This inclusion of exchange effects in (VII,5) often results in a considerable increase of the lifetime of a cluster within nuclear matter over the estimation (VII,4)**.

If the lifetime of the bombarding cluster in the nucleus is prolonged to the extent that it is about the same as, or larger than, the traversal time across the target, then we would expect giant optical resonance phenomena to occur. Under these circumstances an average potential of the general form (VII,1) can be regarded as sufficient for describing these

---

* Because the nucleons are indistinguishable such exchanges are of course unobservable. To assume that eq. (VII,4) correctly estimates the lifetime of an $\alpha$-cluster within a larger nucleus is to assume an initial wave function of the form (VII,5) without the antisymmetrization operator, which means we incorrectly describe the nucleons as distinguishable particles.

** As extreme examples the low-lying states of $O^{16}$ and the low energy $\alpha - \alpha$ scattering states considered earlier have an $\alpha$-cluster structure which persists indefinitely in first approximation.

optical effects. Stated otherwise, when the cluster structure (VII,5) is sufficiently stable, it is reasonable to expect to be able to represent the interaction of the clusters, at least approximately, by means of some energy dependent complex optical potential. The situation would be much the same as with ordinary nucleon-nucleus scattering, as noted previously. The real part of the effective potential between the clusters has a depth corresponding to $V_0$ in (VII,1) of the order of $50-100$ MeV or more$^\star$, and should vary only slowly for small bombarding energies of, say, less than $25-30$ MeV. The imaginary part of this potential, which is also angular momentum dependent, depends much more on the special target nucleus in consideration than the real part. This imaginary part essentially describes the decay probability and the inelastic scattering probability of the bombarding cluster in the target nucleus.

Such energy and angular momentum dependent optical potentials for nucleus-nucleus scattering can be calculated at least in principle with the nuclear reaction theory sketched in chapter V. As pointed out there, the integral kernel which describes the interaction of the clusters (bombarding cluster and target nucleus) and their transitions to other channels is usually non-hermitian. The effect of this kernel on the two clusters can be replaced by a usually complex optical potential (see text to eq. (V,41) and (V,42)). Examples of such theoretically derived optical potentials are the earlier discussed optical potentials of $\alpha - \alpha$ scattering, $\alpha$-He$^3$ scattering and $\alpha$-nuclear matter scattering. Due to the fact that in their calculation we neglected transitions to other channels we obtained real optical potentials, but a more general treatment along the lines of eq. (V,41) would yield complex optical potentials.

The usefulness of such theoretically derived optical potentials will depend on how strongly they vary with bombarding energy. In our example (see for instance the text and footnote following eq. (V,30a)) we have seen that the optical potential for a given incident channel is strongly energy dependent near resonances due to cluster structures in other channels. Therefore we can roughly expect the following behaviour: At low energies the stationary scattering states can often be well approximated by a single cluster structure. In this energy region the derived optical potential will be rather independent of energy, and will produce rather narrow resonances which describe the low-lying virtual levels of the nucleus having this cluster structure. An instance of such optical levels are the $\alpha$-particle virtual states of Be$^8$ discussed earlier (chap. V sect. C subsect. a). Above these low energies an adequate approximate stationary scattering state wave function requires a superposition of cluster functions, the number of terms increasing with the energy and representing both the open and closed channels (see text to eq. (V,15)). The derived optical potential will then become strongly energy dependent due to resonances in these other channels.

$^\star$ This potential depth is roughly $V_0\, n_B$ with the magnitude of $V_0$ in eq. (VII,1) and $n_B$ the number of nucleons in the bombarding particle. In chapter V section D it was calculated explicitly for the $\alpha$-nuclear matter scattering case to be about 120 MeV and slowly increasing with (small) increasing scattering energies [83].

The superposition of many cluster structures required in the stationary scattering state wave functions above the lowest energies essentially accounts for the formation of the compound nucleus at these energies. To see this more descriptively consider a wave packet $\psi_p$ which at $t = 0$ is confined to the nuclear volume and represents just the cluster structure in the incident channel[*].

$$\psi_p\,(t = 0) = \mathscr{A}\phi_B\,\phi_T\,\chi\,(\boldsymbol{R}_B - \boldsymbol{R}_T)\,. \qquad (\text{VII},5\text{a})$$

To build up such a packet $\psi_p$ will require a superposition of many stationary scattering states with some mean energy $E_0$ and a usually rather broad energy spread $\Delta E_p$ so chosen that at $t = 0$ just the cluster structure of (VII,5a) is present in the wave packet. The wave packet now develops in time and evolves away from its initial form in a characteristic time $\tau_p$ defined by

$$|\langle\psi_p\,(t = 0)|\,\psi_p\,(t = \tau_p)\rangle| = \frac{1}{e} \qquad (\text{VII},5\text{b})$$

and related to $\Delta E_p$ by the uncertainty principle $\tau_p \cdot \Delta E_p \geqq \hbar$. The evolution of the packet occurs in two ways: (1) it begins immediately to leak out as the corresponding free clusters, roughly corresponding in the optical model to elastic scattering without compound nucleus formation, and (2) the part of the wave packet remaining in the nuclear volume begins to evolve from its initial cluster structure into a superposition of different cluster structures which start leaking out into the various open reaction channels, corresponding roughly in the optical model to absorption due to compound nucleus formation and decay. The method of cluster representations thus contains in principle a formulation of the compound nucleus behaviour, N. BOHR [115], see also the resonating group formulation of J. WHEELER [67].

The usefulness of wave packets is that we can often make rough estimations of the characteristic times $\tau$ of wave packet processes and connect them with the energy widths $\Delta E$ of resonances in the corresponding cross-section of stationary states by the usual inference:

$$\tau \cdot \Delta E \geqq \hbar\,. \qquad (\text{VII},6)$$

Hence for instance the optical elastic scattered part of the wave packet can only leak out with a significant amplitude in the time $\tau_p$ that the initial cluster structure $\psi_p\,(t = 0)$ survives in the compound nucleus, so that in the stationary state elastic scattering cross-section we expect a

---

[*] In contrast to stationary state wave functions the relative motion function $\chi\,(\boldsymbol{R}_T - \boldsymbol{R}_B)$ of $\psi_p$ in (VII,5a) is taken as a localized wave packet. We emphasize that wave packets and stationary states represent two different incompatible physical situations (the one localized in space, the other in energy). Wave packets such as (VII,5a) cannot usually be realized in practice, but we can often infer from them useful information about stationary scattering states, which are a good approximation to the usual laboratory scattering experiments. For instance the lifetimes discussed earlier [text to (VII,5)] strictly refer to wave packet processes, since they have no meaning in a stationary state, but we can infer resonance widths of stationary state cross-sections from such "gedanken" experiment wave packet processes, see for instance J. BLATT and V. WEISSKOPF [114], p. 386.

rather broad optical resonance with width of the order of magnitude $\Delta E \approx \hbar/\tau_p$. This optical resonance will extend over many narrow compound nucleus resonances whose average width is to an order of magnitude $\Delta E \approx \hbar/\tau$ where $\tau$ is the characteristic time for the wave packet $\psi_p(t)$ to decay completely out of the region of the compound nuclear volume through all channels, see for instance [106].

The theoretically derived optical potential for the incident channel, since in principle it is calculated from as good an approximate stationary scattering state wave function as desired, will exhibit all compound nucleus resonances. Hence it will be strongly energy dependent in this energy region of sharp compound nucleus resonances. This potential may be averaged over the energy and the resulting relatively energy-insensitive average optical potential will be able to produce only the broad optical resonances, see Fig. 24.

It is this averaged optical potential which corresponds to the usual phenomenological optical potentials chosen for the "gross structure" description of nuclear scattering and reaction processes, see for instance [106]. At still higher bombarding energies the compound nucleus resonances begin to overlap so strongly that only this average behaviour is observed, i. e. in Fig. 24 at higher energies the dotted line becomes the observed cross-section.

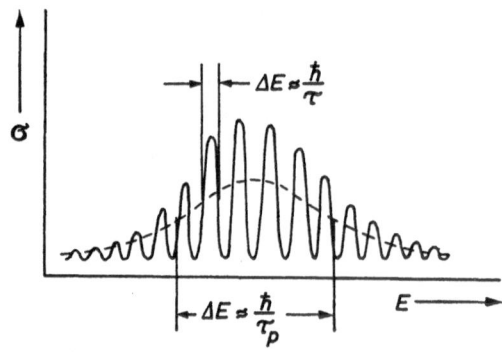

Fig. 24. Schematic elastic scattering cross-section in energy region of sharp compound nucleus levels; solid line: observed, dotted line: averaged

For the purpose of rough estimation it is often more convenient to think of the characteristic time $\tau_p$ determining the width of an optical giant resonance, $\Delta E \approx \hbar/\tau_p$, as arising from two separate processes: corresponding to (2) above, $\tau_{\text{lifetime}}$ is defined as the length of time the bombarding cluster spends in the compound nucleus before it is destroyed or loses its relative motion by exciting the target nucleus; corresponding to (1) above, $\tau_{\text{transit}}$ is defined as the length of time the bombarding cluster spends in the target nucleus (providing it is not destroyed) before leaking out without compound nucleus formation.

$\tau_{\text{transit}}$ may also be defined equivalently as the time required for the bombarding cluster to go from one edge of the target to the other, $\tau_{\text{traversal}}$, multiplied by the number of times it is reflected back at the nuclear surface. For high bombarding energies $\tau_{\text{transit}}$ approaches $\tau_{\text{traversal}}$. Often one of these processes dominates the other, that is, very often either $\tau_{\text{lifetime}}$ is much larger than $\tau_{\text{transit}}$, or $\tau_{\text{lifetime}}$ is so short that the cluster is destroyed before it can make even one traversal of the nucleus. Then $\tau_p$ is either $\approx \tau_{\text{transit}}$ or $\approx \tau_{\text{lifetime}}$.

*b) Examples of optical resonances in the incoming channel*

We consider first the scattering of $\alpha$-particles on $C^{13}$ for medium bombarding energies (5–10 MeV). A rough estimation which we will not describe here shows that $\tau_{\text{lifetime}}$ for an $\alpha$-cluster in $C^{13}$ for such bombarding energies is approximately $2 \cdot 10^{-20}$ sec, [113]. The transit time for such bombarding energies is about $3 \cdot 10^{-22}$ sec (we are well above the Coulomb barrier). Therefore the energy widths for the expected optical giant resonances are given by

$$\Delta E = \frac{\hbar}{\tau_{\text{transit}}} \approx 2\ \text{MeV}. \tag{VII,7}$$

The energy spacing between these giant resonances should be some MeV as inferred from the energy separation of the $\alpha - C^{12}$ cluster states in $O^{16}$ and the $\alpha - C^{13}$ and $\alpha - N^{13}$ cluster states in $O^{17}$ and $F^{17}$ respectively (see for this chap. III sect. E and Fig. 4, [113] and [30]). As yet, $\alpha - C^{13}$ elastic cross-sections have not been measured, but other experimental data are discussed in subsection d) below and chapter VII section C (see also [113]) from which we are able to infer that the qualitative predictions for optical resonance phenomena above are substantially correct.

As next example we consider the scattering of $C^{12}$ on $C^{12}$ and of $O^{16}$ on $O^{16}$, which have been measured, E. ALMQUIST et al. [116]. The remarkable feature of these measurements is that an optical resonance structure is observed in the $C^{12}$-$C^{12}$ scattering but not in the $O^{16}$-$O^{16}$ scattering. It is of considerable interest therefore to attempt an explanation of these findings from our point of view.

Because $C^{12}$-clusters have unfilled outer shells, two $C^{12}$-clusters overlap each other rather strongly. Hence we except a strong mutual interaction between them. From chapter V section D subsection a) we can therefore expect a deep optical well for two $C^{12}$-clusters, with several optical resonances below and near the top of the Coulomb barrier. Also from the strong interaction we expect $\tau_{\text{transit}}$ to be large, which means that $\tau_{\text{lifetime}}$ essentially determines the width of the resonances at these energies. Estimation of the lifetime of two unexcited $C^{12}$-clusters which penetrate each other gives the value $(2-5) \cdot 10^{-21}$ sec, see [113]. The corresponding partial energy width is of the order of 100 KeV. Therefore for the scattering of $C^{12}$ on $C^{12}$ one should expect optical giant resonances. For this scattering we can expect the following results. For incident energies well below the Coulomb barrier all resonance effects will be damped out. The elastic cross-section will be pure Mott scattering and the reaction cross-sections will be very small. At energies just below and above the Coulomb barrier, optical resonance effects should appear in the scattering and therefore in the reaction cross-sections*. In the energy neighbourhood of the top of the Coulomb barrier ($\sim$ 15 MeV c.m. scattering energy) an entrapped cluster undergoes many reflections in the nucleus, and the widths of the resonances should be about 100 KeV as already stated**, because in this case due to a large reflection probability $\tau_{transit}$ is of the order of magnitude of, or larger than, $\tau_{lifetime}$. As the energy of the incident $C^{12}$ particles is increased further, the resonance structure should smear out and eventually disappear. This comes about because of the decreasing transit time. At high scattering energies the $C^{12}$-clusters spend little time near each other ($\sim 5 \cdot 10^{-23}$ sec) as the reflection coefficient at the nuclear surface approaches zero. The description we have presented is in qualitative agreement with the experimentally measured cross-sections, especially with the reaction cross-sections, [116].

The fact that no resonance effects occur in the scattering of $O^{16}$ on $O^{16}$ must be attributed to the smallness of the effective interaction between two $O^{16}$-clusters compared to that of two $C^{12}$-clusters. This can be theoretically understood from the Pauli principle. Nucleons in the $C^{12}$-clusters can penetrate into their mutual unfilled $p$-states, while this can occur in $O^{16}$ only by internally exciting (i.e. breaking up) the $O^{16}$-clusters because they are closed shell clusters. This small overlapping results in a relatively small effective interaction between two proximate $O^{16}$-clusters and favours a large relative kinetic energy. Therefore it is understandable that two unexcited $O^{16}$-clusters do not form a bound or quasi-bound state below their Coulomb barrier (for more details see [113], see also E. VOGT and McMANUS [117] and R. DAVIS [118]). This being the case, no optical resonance phenomena are expected in the scattering of $O^{16}$ on $O^{16}$. That not even broad resonances are observed above the Coulomb barrier indicates that the effective interaction potential between the two $O^{16}$-clusters is rather shallow, as expected.

It is apparent from the nature of our considerations that optical giant resonance phenomena and optical interference phenomena should be very general features of nuclear reactions and should occur in many reactions of heavy nuclei. One should expect optical resonance phenomena for example in the scattering of $\alpha$-particles, tritons, $He^3$ and even deuterons from $O^{16}$. Giant resonances occuring near the top of the Coulomb barrier should have relatively small widths (about some 100 KeV), as mentioned before. At higher scattering energies the resonances should become very broad (in the MeV range) reflecting the decreasing value of $\tau_{transit}$. It is always necessary that if the latter broad giant resonances are to occur, the slope of the effective potential between the clusters must be rather sharp near its edge.

---

* Since the reaction cross-sections depend directly on the capture probability of the incident particle, any resonance behaviour associated with the incoming channel of a nuclear reaction should usually be exhibited in all allowed reaction channels.

** The energy separation of these levels should be about 300—500 KeV, see [113].

### c) Optical resonance phenomena in outgoing reaction channels

Our previous considerations have dealt with optical resonance phenomena only when the resonances were associated directly with the incoming channel of the nuclear reaction. Under certain circumstances however, one should expect optical resonance phenomena to occur in a specific outgoing channel in a nuclear reaction even if resonances do not occur in the capture cross-section. This comes about as follows. Consider a nuclear reaction which is initiated, for example, by the capture of a nucleon with an energy of several MeV by a medium-heavy or heavy nucleus. After the capture of the bombarding particle the highly excited compound nucleus formed goes through all the modes of excitation (described for instance by many different kinds of cluster configurations) allowed by conservation laws*, see for this also the discussion subsection a). For example, suppose that the compound nucleus allows the formation of two excited or unexcited cluster states with a finite probability; these states are described by wave functions of the general form

$$\psi = \mathscr{A} \{\phi_a(\bar{r}_1, \ldots \bar{r}_a) \ \phi_b(\bar{r}_{a+1}, \ldots \bar{r}_A) \ \chi(R_a - R_b)\} \qquad \text{(VII,8)}$$

where the notations are the same as in eq. (VII,5).

If the lifetime of a cluster state such as (VII,8) is of the order of $10^{-21} - 10^{-22}$ seconds or larger, and if the energy for the compound nucleus decaying into the corresponding free nuclei (excited or unexcited) is about equal to their mutual Coulomb barrier, conditions are favorable for an optical giant resonance structure to occur in the energy dependence of the appropriate partial decay cross-section.

A true optical resonance behaviour in the reaction channels can easily be obscured or mistakenly identified. The relevant cross-section is always the ratio of the partial reaction cross-section to the total reaction cross-section, as a function of energy. Otherwise the optical effects we are concerned with will be superposed upon the variation (with incident bombarding energy) of the capture probability of the bombarding particle.

An important qualification for observing true optical resonance phenomena in an outgoing reaction channel is that the probability for the highly excited compound nucleus decaying into particular decay modes must not depend sensitively on the energy of the bombarding particle (which is about 10 MeV or so). Thus the probability $W$ for formation of the cluster structure in the reaction channel will be proportional in first approximation to (using the notation of (VII,5) and (VII,8)):

$$W \propto |\langle \mathscr{A} \ \phi_a \phi_b \chi(R_a - R_b)| \ \bar{H} - E \ |\mathscr{A} \ \phi_B \phi_T \chi(R_B - R_T)\rangle|^2 . \qquad \text{(VII,8a)}$$

This probability will usually be large when the wave function $\mathscr{A} \ \phi_a \phi_b \chi$ in the reaction channel is large in the nuclear volume, which is the case near a resonance in that reaction channel. However it can happen that

---

* The assumption of a bombarding energy of several MeV allows the nucleon to enter the nucleus through many incoming channels, especially if the energy spread of the beam is not too small. It also means that many outgoing channels are open for the subsequent decay of the compound nucleus.

at the bombarding energy in the incident channel corresponding to this resonance in the reaction channel the matrix element is nevertheless small if the right hand side $(\overline{H} - E) \mathscr{A} \phi_B \phi_T \chi$ is markedly energy dependent. Then all resonances would be a complicated superposition of optical effects in the reaction channel plus this energy dependence. This difficulty should not usually arise. Since the effective interaction energy between two clusters is usually much larger than the bombarding energy of the captured particle (say 50 MeV and more compared to about 10 MeV)*, the energy dependence of the probability for forming a pair of clusters** should be essentially independent of the right hand side of the matrix element (VII,8a) if no optical resonances occur in the incoming channel over the bombarding energy range considered. Thus we can expect that, at least to a first approximation, the structure of giant resonances in partial reaction cross-sections should be independent of the manner of formation of the compound nucleus. Furthermore the giant resonances in a particular reaction channel should be very similar to the giant resonances occuring in the reverse elastic scattering process, excluding Coulomb effects, when that channel is made the incident channel.

It is very difficult to do more than estimate very roughly the order of magnitudes of partial reaction cross-sections, because this would require knowing the reaction mechanism for the creation of the different cluster structures allowed. That is, one would have to know the time development of the compound nucleus system starting from the capture of the bombarding particle***. One can only expect that these partial cross-sections are much smaller than the corresponding total cross-section $\left(\text{say by a factor of } \frac{1}{20} \text{ or less}\right)$ and much larger than what any statistical model would predict. The latter is essentially due to the fact that the Pauli principle greatly reduces the differences between different cluster structures, so that very often the compound nucleus can pass rather quickly from one cluster structure to another merely by the rearrangement of a few nucleons.

### d) Examples of optical resonances in reaction channels

In the $O^{16}$ (ground state) (p, $\alpha$) $N^{13}$ (ground state) reaction for proton energies from 6 to 20 MeV it is observed that the partial cross-section displays optical giant resonances with width of $1-2$ MeV†, WHITEHEAD and FOSTER [119], R. SHERR et al. [120]. The peak of the first giant resonance occurs for a proton energy of about 8.5 MeV corresponding to a decay energy of 3 MeV.

---

* For example the interaction energy of two not internally excited $C^{12}$-clusters is about 50—100 MeV (see [113]).

** In the nuclear reaction theory sketched in chapter V this probability depends on all transition kernels to the particular channel of the clusters considered.

*** In principle this can certainly be calculated from the associated (time-dependent) set of coupled channel equations.

† The total p-$O^{16}$ cross-section has a smooth behaviour in this energy region.

As discussed in subsection c), the same optical resonance structure should appear in the appropriate reverse scattering process and its mirror reaction. Hence the observed optical resonance in the $\alpha$-$N^{13}$ reaction channel above tends to confirm our remarks in subsection b) for an expected optical resonance in $\alpha$-$N^{13}$ and $\alpha$-$C^{13}$ elastic scattering.

It would be very interesting to investigate experimentally the extent to which the structure of the giant resonances associated with the outgoing channels of a nuclear reaction is really independent of the way in which the compound nucleus system is formed. To this end one could investigate such reactions as: $N^{15}$ (d, $\alpha$) $C^{13}$ (ground state); $N^{14}$ (t, $\alpha$) $C^{13}$ (ground state); etc.

As our next example we discuss briefly the experimental results of the $Al^{27}$ (n, $\alpha$) $Na^{24}$ reaction, H. SCHMITT and J. HALPERIN [121] for neutron energies from 6.1 to 8.3 MeV*. An optical giant resonance structure is observed with widths roughly about 100 KeV and energy separations between resonances of about 0.5 MeV. This is expected when one notes that the Coulomb barrier for $\alpha$-$Na^{24}$ scattering is about $4-5$ MeV. Thus the energy in the $\alpha$-$Na^{24}$ decay channel in the above experiment extends from a little below to just above the Coulomb barrier. In this reaction it is the lifetime of the $\alpha$-cluster in the $Na^{24}$-cluster (without exciting the $Na^{24}$-cluster) and not its transit time that determines the widths of the giant resonances. In other words the cluster reflects several times but decays within the nucleus or excites it before it leaks out appreciably. We have mentioned in b) that $\tau_{lifetime}$ of an $\alpha$-cluster in a light nucleus should be $(2-5) \cdot 10^{-21}$ sec, corresponding to an energy width of some 100 KeV.

In experiments if one is not careful to separate the open channels carefully, an optical resonance in one channel may be washed out by the presence of other open channels. Therefore experimentally observed giant resonances would probably be best exhibited by measuring decays to a specific state of the residual nucleus (i.e. one specific channel), for example the ground state of $Na^{24}$.

In ending this section concerning optical giant resonances in nuclear reactions, we want to emphasize again that the Pauli principle (indistinguishability of the nucleons) is responsible for the fact that clusters in nuclear matter can exist sufficiently long without decaying that optical giant resonance features can appear in certain nuclear reactions. We repeat again, this does not mean that during the mutual penetration of the clusters their nucleons cannot be exchanged. On the contrary the exchange-time for two nucleons in different clusters (for instance in $\alpha$-$\alpha$ scattering) during the mutual penetration of the clusters is about $10^{-23}-10^{-24}$ seconds. But these exchange effects are automatically taken into account by the antisymmetrization of the total wave functions. If the mutual penetration of the clusters decreases, certainly the exchange time increases greatly. For a more detailed treatment of these optical giant resonance considerations see [113].

---

* The decay energies range from 3—5 MeV, providing $Na^{24}$ is not excited after the decay.

## B. Nuclear fission

We now apply the cluster viewpoint to understand the phenomenon of asymmetric fission in heavy nuclei. A partially successful theory of nuclear fission has been available for some time, namely that provided by the liquid drop model, N. BOHR and J. WHEELER [122], J. FRENKEL [123]. In this model one envisions a heavy nucleus, when excited (as for instance by the capture of a thermal neutron), as deforming until it splits into two parts. The key effect which lowers the energy of the intermediate deformed state enough to give an observable fission rate for nuclei of large atomic weight $A_0$ is the cancellation of the increased surface energy by a decrease in the Coulomb energy. Using this picture

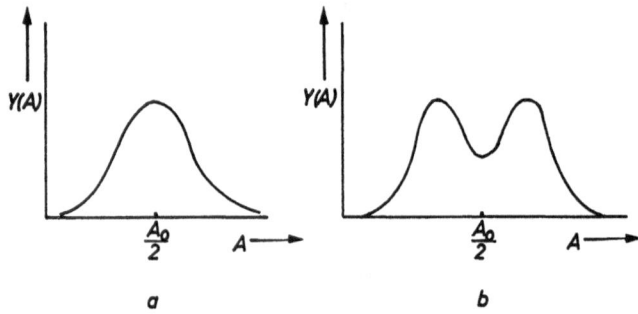

Fig. 25 a and b. Schematic mass distribution of fission products. a) liquid drop model, b) asymmetric fission

of the fission process one can make some hydrodynamical calculations and predict a "symmetric" curve (sketched in Fig. 25a) for the distribution in atomic weight $A$ of the spallation products, which has a peak at half the atomic weight $A_0$ of the starting compound nucleus.

Experimentally we obtain results in general agreement with this curve for induced fission of some nuclei. However, in contradiction to this model we observe in other cases a curve with two peaks located symmetrically about $\frac{A_0}{2}$, known as the asymmetric fission curve, shown by the curve in Fig. 25b*. The asymmetric distribution is observed for spontaneous fission or induced fission of nuclei with neutron number $N \geq 132$ at sufficiently low energies of the bombarding particle. The question arises as to how we are to understand this asymmetric curve.

Since the peaks in the asymmetric cases correspond to one fission product with $N = 82$ or slightly more and the other with $N = 50$ or slightly more and since asymmetric fission occurs only in nuclei with at least $80 + 50 = 132$ neutrons, it is easy to conclude that it has something to do with the stability of the magic number configurations of 50 and 82 neutrons, L. MEITNER [124, 125]. But it is not obvious how magic number effects fit into the liquid drop picture. One widely discussed

---

* Any distribution of fission products must necessarily be symmetric about $\frac{A_0}{2}$ if we assume that each nucleus splits into only two pieces.

attempt is to argue that magic nucleides have a lower mass than neighbouring nucleides, and as the total energy is constant, presumably they are formed with a higher excitation energy than the non-magic nuclei at the moment of scission. Because the level density rises rapidly with excitation energy the phase space and therefore the transition probability for the formation of magic nuclei with a few outside nucleons should be greatly enchanced, if one assumes the transition matrix element is essentially constant, P. FONG [126]. But experiment shows that magic number fragments tend to come off with a lower mean excitation energy than symmetric fragments, in contradiction to the above assumption. To get around this and other difficulties appears to require additional ad hoc postulates. Furthermore theoretical considerations indicate that the transition matrix elements are not constant and depend very much on the final state configuration. Therefore one cannot neglect the role of magic number effects (closed shell effects) in determining the magnitude of the matrix elements themselves.

The preceding discussion shows that one has to add essentially new features to the classic explanation of nuclear fission in order to understand asymmetric fission. On the other hand we only wish to refine the liquid drop model and not invalidate it completely, because it provides a very natural basis for the explanation of many important features of nuclear fission*. The refinement we seek is to explicitly consider the effects of cluster substructures, and particularly the behaviour of closed shell clusters**, within the liquid drop model, H. FAISSNER and K. WILDERMUTH [127].

To begin our discussion from the cluster viewpoint, consider the case of induced fission where the original nucleus absorbs a slow neutron to form a fission prone intermediate (compound) nucleus with $N \geq 132$. Then in effect the nucleus is excited to the neutron binding energy of $6-8$ MeV***.

Now any wave function may be represented by a superposition of two cluster wave functions, since they comprise a complete set (see chapter IV sect. B and chapter V sect. B). Hence we may write the (time-dependent) wave function of the excited intermediate nucleus as

$$\psi(t) = \sum_{N_1, Z_1} \sum_{i, j, k} a_{i, j, k}^{N_1, Z_1}(t) \, \mathscr{A} \, \phi_i(N_1, Z_1) \, \phi_j(N_2, Z_2) \, \chi_k(\boldsymbol{R}_1 - \boldsymbol{R}_2) \quad \text{(VII,9)}$$

where $\phi_i$ and $\phi_j$ represent the $i^{\text{th}}$ and $j^{\text{th}}$ internal state of the clusters $1$ and $2$ respectively and $\chi_k$ the $k^{\text{th}}$ state of their collective relative motion. Note $N_2 = N_0 - N_1$, $Z_2 = Z_0 - Z_1$. The total time dependence of $\psi(t)$ is contained in the amplitudes $a_{i, j, k}^{N_1, Z_1}(t)$. Although the two-cluster states in the superposition (VII,9) are not all orthogonal, they do not overlap in

---

* For instance it explains why no stable nuclei with atomic weight $A$ larger than about 250 can exist.

** In this section the phrase closed shell clusters refers to clusters with just a few nucleons outside a closed shell.

*** In slow neutron fission the intermediate nucleus must contain an even number of neutrons because in odd $N$ nuclei the pairing effect reduces the binding energy of the odd neutron below the threshold excitation for fission, [122].

the asymptotic region of large cluster separation so there is no problem defining cross-sections (see chapter V sect. B). We may think of the nucleus as passing simultaneously through many different kinds of cluster configurations, symmetric and asymmetric, as the coefficients $a_{i,j,k}^{N_1,Z_1}(t)$ develop in time.

De-excitation of the excited intermediate nucleus may now occur in only very few ways: in addition to the fission modes, gamma emission and evaporation of a nucleon are the only decay modes*. The usefulness of the two-cluster expansion (VII,9), which is so far an arbitrary mathematical device, rests in that we can examine the specific terms which contribute to a particular de-excitation process. In fact we will show that at low energies only those terms representing asymmetric closed shell cluster configurations contribute significantly to de-excitation by fission.

Let us review some characteristics of closed shell clusters. In the arguments of chapter IV section D we saw that free clusters which are especially tightly bound, e.g. clusters with filled shells $N = 50$, $N = 82$, or $Z = 50$, etc., remain relatively especially tightly bound** even when surrounded by many other nucleons. This tight binding of closed shell clusters plays a crucial role in low energy asymmetric fission.

For a long time it was not understood how a heavy nucleus ($N \geq 132$) could have especially tightly bound substructures with $N = 50$ and $N = 82$ without a complete break-up of the $N = 126$ and other closed shells of the nucleus, which would more than offset the energy gained by closing the shells of the substructures. But antisymmetrization (i.e. the Pauli principle) so strongly reduces the differences between different cluster structures that this difficulty is resolved.

To make this last point in more mathematical terms, consider a closed shell configuration of $U^{236}$, say, in the frame of the oscillator assumption: $\psi = \mathscr{A} \, \phi(1) \, \phi(2) \, \chi(R_1 - R_2)$. (For instance $\phi(1)$ might represent a cluster with $Z_1 = 51$, $N_1 = 83$ in which the $Z = 50$ and $N = 82$ shells of the cluster are filled and the few odd nucleons lie in the next shells of the cluster, and $\phi(2)$ a cluster with $Z_2 = 41$, $N_2 = 51$ in which the $Z = 40$ and $N = 50$ shells are filled.) If we were to expand this oscillator cluster function in oscillator shell model functions, then we would expect to find the single particle shells $Z = 82$, $N = 126$ completely filled and the remaining nucleons in some superposition of single-particle levels of the next shell such that the clusters are especially tightly

---

* Both of these latter modes are long lived however; $\tau_\gamma$ and $\tau_{evap}$ are of the order of magnitude of $10^{-12}$ to $10^{-15}$ sec. $\tau_{evap}$ is large because one nucleon must regain practically all the "heat" energy before it can escape again, which is very improbable. Neutron evaporation occurs when the nucleus passes through a cluster configuration with $N_1 = 1$, $Z_1 = 0$; $N_2 = N_0 - 1$, $Z_2 = Z_0$ with the residual nucleus unexcited, [122].

** It is perhaps more correct to refer to this energy as the correlation energy associated with a particular cluster configuration rather than as a binding energy, for the binding energy of a cluster within a larger nucleus is somewhat artificially defined, see eq. (V,6a).

bound (see chap. IV sect. D)*. If we leave the oscillator assumption and use generalized cluster functions, the above conclusions still remain very nearly correct.

We are now ready to discuss the role of closed shell clusters in asymmetric fission. For the sake of clarity we will sketch the principal argument at first in the language of the optical potential section. Once the main idea is clear we will refine the argument.

The principal argument may be stated as follows. As the amplitudes $a_{i,j,k}^{N_1,Z_1}(t)$ in eq. (VII,9) vary in time the intermediate (compound) nucleus passes through many cluster configurations of the same (within the limitation of the uncertainty principle) energy. Statistically it is clear that it will spend most of its time in configurations for which the two clusters have nearly equal mass. (More accurately it will be found with a higher probability in such configurations.) With a lesser probability it will pass through configurations in which the two clusters have unequal mass. However these various cluster configurations are not all equally fission prone because they are not all equally deformable. A deformation of the intermediate nucleus may be regarded as simply an increased separation between the two ultimate clusters composing it at any instant. From the liquid drop model we know that the more these two clusters can separate, the lower will be the final potential barrier (usually called Coulomb barrier) holding them together and hence the greater will be the probability for the clusters to tunnel out free of each other. But the separation between the clusters can be especially large only if the energy (kinetic and potential) in the relative motion of the clusters about each other is especially large. Because the intermediate nucleus has a virtually fixed total energy, this large relative energy can only be achieved at the expense of the internal (or binding) energies of the two clusters, $E_{REL} = E_{TOT} - E_{INT}$, see eq. (V,6a). Thus at low excitation energies ($E_{TOT}$ small) $E_{REL}$ may be high only if $E_{INT}$ is especially large and negative, i.e. if the clusters are especially tightly bound. This is indicated schematically in Fig. 26. In the figure $E_{REL}$ (sym) denotes relative energy available to nearly symmetric clusters, $E_{REL}$ (asym) denotes relative energy available to clusters with closed shells, $V_{opt}$ denotes the effective (optical) potential holding the clusters together; this potential will be somewhat different for different cluster structures.

Perhaps more accurately we may say that the energy released by correlating the nucleons so as to achieve especially tightly bound clusters is available for the collective energy of deformation. Hence the intermediate nucleus is particularly fission prone only in those asymmetric

---

* How this comes about may also be seen in Appendix C where we treat Li⁶, which is small enough for the expansion to be carried out explicitly. In Li⁶ we fill the $N = 2$, $Z = 2$ shells (1s shell) of the α-cluster and put a neutron and proton respectively in the $N = 2$, $Z = 2$ shells (1s shell) of the deuteron cluster. In the expansion the $N = 2$, $Z = 2$ single-particle levels of the shell model remain filled and the outside nucleons are coupled to a spin one configuration so as to make the two clusters especially tightly bound. A second example is Be⁸, discussed repeatedly in earlier chapters.

cluster configurations for which the two ultimate clusters have unbroken closed shells.

The probability that the intermediate nucleus possesses the rather highly ordered closed shell cluster structure leading to asymmetric fission is of roughly the same order of magnitude as the probability that

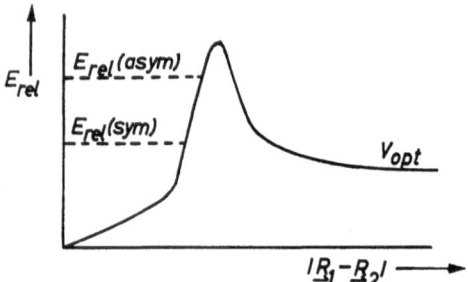

Fig. 26. Heuristic plot of relative energy distribution vs. cluster separation distance

one neutron regains all the excitation energy. Hence the asymmetric fission process and neutron evaporation compete in heavy nuclei if enough energy is available for either. Because no decay channels are open to the intermediate nucleus in its more probable configurations, it must simply wait until it passes through one of these less probable configurations before it can decay*.

We see from this discussion that the relative height of the final potential barrier plays a decisive role in determining the peak to valley ratios in the mass distribution. The sensitivity of the quantum mechanical tunnel effect to the height and width of this potential barrier makes it immediately understandable why small differences in the deformation energy (i.e. relative energy) can produce such large differences in the decay probabilities of the fission products (factor $400-1000$ for low energy fission). Certainly the probability that the intermediate nucleus passes through a configuration in which the two clusters have almost the same proton and neutron number is greater than for an asymmetric closed shell cluster configuration. But in clusters of almost equal mass we must break up the $Z = 50$, $N = 82$ shells of the larger cluster; consequently symmetric cluster configurations have much less correlation energy and the deformation energy cannot be so great (i.e. $E_{\text{INT}}$ is less negative hence $E_{\text{REL}}$ cannot be so large). Thus the potential barrier for the separation of nearly symmetric clusters is usually so high, at low energies, that symmetric fission will occur very rarely compared to asymmetric fission. At the end of this section we check by an order of magnitude estimate that the substructure effects described above are large enough to produce threshold differences corresponding to the experimental peak to valley ratios in the mass distribution.

* The wave function (VII,9) has such a narrow spread in energy that many different cluster structures will be present simultaneously. Therefore one should understand our picturesque language above in the same light by which we connected time-developing wave packets with stationary states in chapter VII section A subsect. a).

We should point out that the form of the (optical) potential preventing
the two clusters from separating will be different for closed shell clusters
than for nearly symmetric clusters. For we recall in chap. V sect. D that,
for instance, the optical potential for $C^{12} - C^{12}$ scattering, which are
nuclei rather far from closed shells, was deeper than the optical potential
for $O^{16} - O^{16}$ scattering, which are closed shell nuclei. Qualitatively this
stronger potential between nonclosed shell clusters arises from the
greater overlapping of the many outside nucleons of the clusters. These
differences should not alter our above considerations.

With increasing excitation energy of the intermediate nucleus the
influence of the potential barrier on the fission process certainly decreases.
This gives a natural explanation for the observed fact that with increasing excitation energy of the intermediate nucleus the relative
probability for symmetric fission compared with asymmetric fission
increases rather rapidly,* see for instance I. HALPERN [128] and E. HYDE
[129]. On the other hand it is clear from this energy dependence that
spontaneous fission, as an extreme "low excitation energy" limit of
induced fission, should exhibit cluster effects most strongly. This is
indeed the case. Contributions from outside the mass bands corresponding to unbroken closed shell cluster configurations are exceedingly small,
the symmetric mode is virtually absent. It is also experimentally confirmed that no asymmetric fission occurs if the fissioning nucleus has
less than 132 neutrons, A. FAIRHALL et al. [130], which is expected since
the closed shell clusters described above cannot be formed. Since the
decision as to how the intermediate nucleus will fission at some given

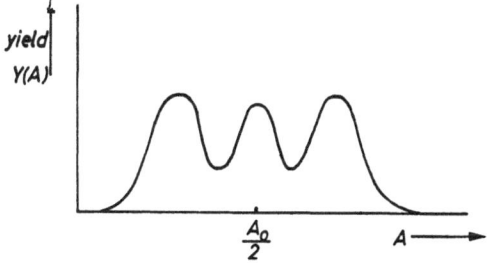

Fig. 27. Schematic mass distribution for Ra²²⁶ fission

excitation energy is thus made at a comparatively early stage of the
fission process, i. e. at small deformations, symmetric and asymmetric
fission should be considered as two different processes [130] in the same
way as, for instance, neutron evaporation is different from the fission
processes. In fact at intermediate energies (bombardment by 11 MeV
protons) both modes have been observed to show up distinctly in Ra²²⁶
as indicated in Fig. 27 [130].

---

* But it does not increase as rapidly as one would have to expect if the total
increase in the excitation energy would be completely transformed into deformation
energy, see for this the end of this section.

With the cluster viewpoint developed above we have now shown the key effect responsible for low energy asymmetric fission. Using this viewpoint we may go ahead and outline the remainder of the fission process.

The part of the compound nucleus wave function, eq. (V,9), describing the saddle-point deformation is characterized by great simplicity, that is, only the relatively few terms representing "cold" closed shell clusters contribute to this part. At low energies there is little energy available at the saddle point for anything but the potential energy of deformation. Consequently the clusters themselves must be rather "cold", and the collective kinetic energy (i. e. in approximate terms, the kinetic energy in the relative motion) is necessarily small. Because of this small kinetic energy the collective motion is necessarily a rotation rather than a vibration. Thus the cluster viewpoint also provides a natural basis for the postulate of a collective rotation of the intermediate nucleus near threshold, A. BOHR [13].

It should be emphasized that the time spent in this state of ordered rotation is much smaller than the lifetime of the intermediate nucleus ($\tau \sim 10^{-15}$ sec). Simple estimates indicate it to be about $10^{-18} - 10^{-17}$ sec, i. e. only a few rotation periods. But this is long enough to produce the striking anisotropy effect on the angular distribution of the fission fragments [130], R. HENKEL and J. BROLLEY [133], see also [128].

The final state of the fission process, the scission, is the descent from the saddle point until the actual separation of the fragments is achieved. Its duration must be short compared to the time spent in a rotational state near threshold. Otherwise angular momentum would be transferred from collective rotation to disordered single particle motion and the anisotropic angular distribution would be destroyed. Extreme deformations happen during this short time, which might be described by the classical model of a pear-shaped drop getting longer and longer. This has been done by many authors, S. WHETSTONE [133], V. VLADI-MIRSKI [134], G. HERMANN [135].

The "neck" between the two separating clusters is primarily formed from the more loosely bound nucleons outside the closed shells of the clusters. These outside nucleons are so distributed between the two clusters that the neutron to proton ratios of the clusters are as nearly as possible the same as the parent intermediate nucleus:

$$\frac{Z_1}{N_1} \approx \frac{Z_2}{N_2} \approx \frac{Z_0}{N_0}.$$

In this way the asymmetry energy is kept to a minimum\*. Immediately after the scission these outside nucleons originally composing the neck are now distended, so that the fragments are strongly deformed, R. VANDENBOSCH [138], H. SCHMITT et al. [139], and will carry out strong surface oscillations which should be accompanied by multipole radiation, G. TEMMER [140]. These surface oscillations will be gradually

---

\* The asymmetry energy is a consequence of the Pauli principle and the nuclear two-body forces, see for instance F. SCÖG [136] and K. WILDERMUTH [137].

transformed into unordered compound nucleus excitation of the fission products (thereby certainly exciting the outer closed shells of the fragments). Most of this excitation energy will be lost in neutron emission, which occurs at this time to relieve the considerable neutron excess of the (prompt yield) fragments.

We should point out that scission need not necessarily happen near the center of the "neck" joining the two unequal clusters [133], [134], [135]. One can imagine scission taking place quite close, say, to the heavy cluster. In this case the masses of the two fragments may be roughly equal; but the original asymmetry of the saddle point deformation shows up in an asymmetry in excitation energy; the almost undeformed large fragment winding up with much less excitation energy than the smaller one, which is now strongly distended by the loosely bound nucleons originally in the neck. Similarly in the opposite case, if the scission occurs close to the light cluster, it will have relatively few loosely bound nucleons and hence little distortion and the heavier cluster will wind up with a higher excitation. This means that scission near a tightly bound cluster is indicated by a minimum in the excitation energy of the corresponding fragment.

On the other hand in symmetric fission the fragments are always quite far from closed shells. The many outside nucleons make these fragments much more deformable than closed shell products. Hence the energy in surface deformation at the moment of scission will be much greater for symmetric fragments than for closed shell fragments, with the consequence that the products of symmetric fission should show a higher final mean excitation energy than those of asymmetric fission, W. STEIN and S. WHETSTONE [141], W. GIBSON et al. [143], J. MILTON and J. FRASER [143] and [139].

Another consequence of these more loosely bound nucleons outside the closed shell clusters is the occurrence of single particle effects. In particular, the long known depression of spontaneous fission in odd-mass nuclides has been interpreted as an effect of the odd nucleon's orbit, [see also footnote preceding eq. (VII,9)]. The high orbital motion of this nucleon tends to exert a pull on the surface of the deformed nucleus in such a way as to inhibit further deformation. For a quantitative account of single particle effects in spontaneous fission see S. JOHANNSEN [144].

We now wish to show the cluster viewpoint can be used to make more refined predictions about the general form of the mass distribution curves of the fission products by taking explicit account of the especially fission prone closed shell cluster configurations described above. The mass distributions for asymmetric fission can be calculated approximately under the following assumptions [127]:

(A) An appreciable contribution comes only from compound nucleus states with unbroken closed shell cluster configurations.

(B) The fission probability is independent of the distribution of surplus nucleons except that the $N/Z$ ratios in the fission products should be approximately equal to $N_0/Z_0$.

(C) To both sides of the allowed mass bands defined by (A) and (B) the probability for asymmetric fission decreases like a Gaussian.

The three assumptions rest on very different footings. The first one is the basic idea of our considerations, as discussed before. Assumption (B) is a plausible conjecture for the gross structure of the distributions*. Assumption (C) is just a mathematically convenient way to express the rapid decrease of the fission probability as soon as the energetically favoured cluster structures are broken up.

The main problem is the specification of the proper cluster states. Both neutron and proton shells have to be considered. The asymmetry energy imposes severe restrictions on the possible combinations of magic proton and neutron numbers. Any neutron to proton ratio differing appreciably from the value of the fissioning parent nucleus $N_0/Z_0$ would mean an energetically unfavourable charge polarization of nuclear matter. The nuclei exhibiting asymmetric fission have $(N_0/Z_0) \approx 1.56$. The only combination of relevant magic numbers giving $N/Z$ close to this value is $N = 82$ and $Z = 50$. Consequently the heavy fragment should always be dominated by this structure. For the light fragment the possibilities are $N = 50$ and $Z = 40$; but these neutron and proton numbers cannot be combined without deviating drastically from $N_0/Z_0$. Due to the high stability of $N = 50$ this shell will only very rarely be broken up in the light clusters. If $Z_0$ and $N_0$ are sufficiently large there will also be some favouring not to break up the $Z = 40$ shell of the light cluster. These considerations may be summarized in the following predictions, which have been experimentally verified:

The heavy fragment's mass distribution has its low mass shoulder at $A_h = 134$, $N_h = 82$, $Z_h = 52$ for all heavy nuclei. Correspondingly, the high mass shoulder of all primary light fragments is at $A_0 - 134$** ***.

The width of the mass peaks varies with $A_0$ and $Z_0$.

More quantitative predictions are possible if one uses the information available from the measured excitation energies $E^*$ of the fragments [142]. The observed sharp minima in $E^*$ at certain mass numbers as pointed out earlier are a direct indication of the relevant tightly bound cluster structures. In thermal neutron induced fission of $O^{235}$ the following mass numbers of the light fragment seem to be distinguished: $A_l = 90,96$ and $100$. In all these numbers the shell $N = 50$ in the light fragment is not broken up. This must be expected because the $N = 50$

---

* In the classical model, S. WHETSTONE [133], V. VLADIMIRSKI [134], where the surplus nucleons are pictured to form the "neck" joining the two separating fragment cores, this corresponds to equal scission probability anywhere along the "neck".

** See for experimental verification P. ARMBRUSTER [145].

*** One can show easily that for this combination of neutron and proton numbers the closed shell structures $N_h = 82$, $Z_h = 50$ are not broken up and, at the same time, the neutron to proton ratio in both the heavy and light fragments differs as little as possible from the neutron to proton ratio in the intermediate parent nucleus. Further both the heavy and the light cluster are even-even clusters. Such clusters are additionally energetically favoured due to the pairing energy which is also a consequence of the two-nucleon forces, see for instance M. GOEPPERT-MAYER and H. JENSEN [146].

shell is much more stable than the $Z = 40$ and $N = 40$ shells*. The measurement of the proton distribution in the fission products shows further that the $N/Z$ ratio in these fragments is very nearly the same as the $N_0/Z_0$ ratio of the intermediate nucleus, see for instance [135]. This is in agreement with above presented asymmetry energy arguments.

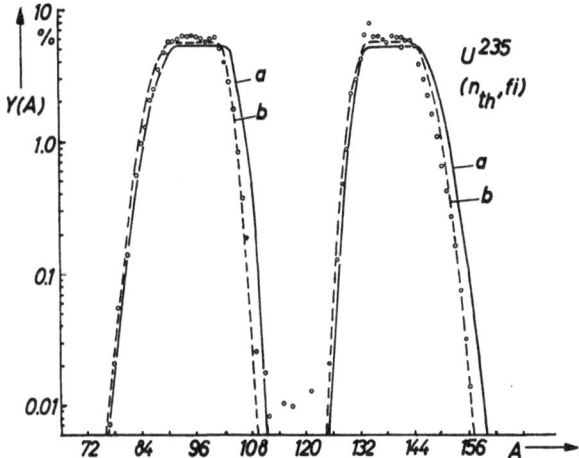

Fig. 28. Mass distribution for thermal neutron induced U²³⁵ fission; $a$ = theoretical prompt yield, $b$ = same corrected for neutron emission, circles = experimental points

Fig. 28 shows an attempt to calculate the thermal neutron fission mass distribution of U²³⁵ under the assumptions (A), (B), (C), using only $A_l = 90$ and $A_h = 132$ as favoured cluster numbers [127]**. The Gaussians have been fitted to the experimental points starting in each case 2 mass units to the *inside* of the shoulders fixed by the cluster numbers, in order to fulfill postulate A. The height of the approximately trapezoidal peak is determined by the normalization. The primary yield curve obtained (curve a) has been corrected for neutron emission using experimental values*** (curve b). The agreement with the experimental points is quite good, although the systematic deviations in the outer wings seem to indicate an influence of the light cluster number $A_l = 84$. This is to be expected due to the remarks in the footnote above.

It seems possible to obtain a good fit to all the observed asymmetric mass distributions by using the relative contributions $\eta(A_l)$ of the different light cluster numbers $A_l$ as adjustable parameters. Unknown distributions can be predicted from the smooth dependence of the

---

* If one assumes $N = 50$ itself as a favoured neutron number of the light (even-even) cluster then it follows that $A_l = 82$ or 84 is also a favoured atomic number for the light cluster provided no appreciable charge polarization happens during the fission process.

** If one would use $A = 134$ as indicated above instead of $A_h = 132$ as favoured cluster number the agreement between calculated and experimental mass distribution would become even better,

*** For more detailed information about the thermal neutron fission mass distribution of U²³⁵ see [127].

$\eta(A_i)$ on $A_0$ and $Z_0$ [127]. An individual adjustment of the slopes is not necessary. All distributions appear to be superpositions of symmetric trapezoids with the slope as given by the inner wings of curve $a$ in Fig. 28. It is interesting to note that for nuclei with $A_0 > 264$ ($N_0 > 164, Z_0 > 100$) the mass distribution is expected to become symmetric with a width $\Delta A \approx A_0 - 264$.

In further extending these considerations one can also conclude that the angular distribution of the fission products with respect to the direction of the incoming neutron beam should be more anisotropic for asymmetric fission than for symmetric fission if the spins of the target nuclei are aligned [127].

We come now to the question raised earlier: are the energetical sub-structure effects described in this section large enough to produce the threshold differences corresponding to the experimental peak to valley ratios in the mass distribution? We check this question by an order of magnitude estimation. First we estimate the cluster correlation energy liberated, and hence available for the energy of deformation, when for a given intermediate nucleus we pass from a nearly symmetrical cluster configuration to an asymmetric closed shell configuration (said in the optical potential language used earlier, we want to estimate the change in the internal energy of the clusters for the two different configurations). To make the estimation we may visualize successively transferring nucleons from the lighter cluster to the heavier cluster. The greatest gain in correlation energy will certainly occur when we transfer the proton and neutron which close the $Z = 50$ and $N = 82$ shells respectively of the heavy cluster*.

As discussed in chapter IV section D, the correlation energy liberated in closing these shells will be smaller but of the same order of magnitude as the corresponding shell effects in free nuclei. Shell effects in free nuclei may be measured by the difference in separation energy, $S(k)$, of a nucleon closing a shell of $k$ nucleons and the separation energy of the $(k + 2)$nd nucleon

$$\Delta(k) \equiv S(k) - S(k+2). \qquad (\text{VII},10)$$

The empirical values for odd mass nuclei are, N. ZELDES [164],

$$\Delta(N = 50) = 3 \text{ MeV}, \ \Delta(N = 82) = 2 \text{ MeV}, \ \Delta(Z = 50) = 2{,}4 \text{ MeV}.$$
$$(\text{VII},11)$$

If we remember that we also gain some net correlation energy in the other nucleon transfers besides the ones which actually close the shells of the heavy cluster, then we may certainly accept $\Delta(N = 82) + \Delta(Z = 50) \approx 4{,}4$ MeV in (VII, 11) as a rough order of magnitude estimate of the liberated correlation energy available for deformation in the asymmetric closed shell cluster configuration.

We now want to find out if this liberated correlation energy in asymmetric closed shell cluster configurations, as roughly estimated in

---

* We emphasize that the $N = 50$ shell of the lighter cluster is never broken up in going from symmetric to asymmetric closed shell clusters for heavy nuclei with $N \geq 132$.

(VII,11), is sufficient to supply the deformation energy required to explain the observed peak to valley ratios of the mass distribution. An estimate of the threshold difference required to produce the observed peak to valley ratio may be obtained by an estimate of the Gamow factor for symmetric and asymmetric fission.

For this estimate we make the following assumptions:

a) The Gamow factor for the most probable asymmetric scission is zero. This means the clusters separate from each other without tunneling through a potential barrier. This assumption is approximately correct for induced fission at threshold.

b) The effective (optical) potential ($V_{opt}$ of Fig. 26) between the two separating clusters is the same for symmetric and asymmetric cluster configurations, and we approximate the nuclear part of this interaction potential by an infinitely steep potential well of radius $R_1$. This is certainly a very crude assumption, and is one essential reason among others that we shall obtain only an order of magnitude estimate.

With assumption (b) we obtain for the Gamow factor of symmetric scission

$$G_{sym} = \frac{2}{\hbar} \int\limits_{R_1}^{R_2} (P_{rel}) \, dR \approx \frac{2}{\hbar} \sqrt{2\mu_{sym}} \int\limits_{R_1}^{R_2} \sqrt{\left| E_{rel} - \frac{Z_0^2 \, e^2}{4 R_1} \left(2 - \frac{R}{R_1}\right)\right|} \, dR \tag{VII,12}$$

with

$$P_{rel}(R_2) = 0 , \quad \mu_{sym} = \frac{A_0}{4} M_{nuc} , \quad E_{rel} = \frac{Z_0^2 \, e^2}{4 R_1} - \Delta E . \tag{VII,13}$$

The integration yields

$$G_{sym} \approx \frac{16}{3 \hbar} \frac{R_1^2}{Z_0^2 \, e^2} \sqrt{\frac{1}{2} A_0 M_{nuc}} \, (\Delta E)^{3/2} . \tag{VII,14}$$

With assumption (a) and neglecting phase space factors* one obtains for the peak to valley ratio near threshold

$$\frac{\Gamma_{asym}}{\Gamma_{sym}} \approx e^{G_{sym}} . \tag{VII,15}$$

From the kinetic energies of fission fragments from $_{92}U^{235}$ ($n_{th}$, fi) (symmetric scission), $E_{rel} \approx 140$ MeV [138], [139] one obtains for $R_1$

$$R_1 \approx 22 \text{ fm} \tag{VII,16}$$

provided one assumes that $E_{rel}$ is equal to $Z_0^2 \, e^2/4 R_1$, which is certainly approximately correct. The experimental value

$$\frac{\Gamma_{asym}}{\Gamma_{sym}} \approx 400 \text{ leads to } G_{sym} \approx 6 . \tag{VII,17}$$

With this we obtain

$$\Delta E \approx 6 \text{ MeV.} \tag{VII,18}$$

This agrees in order of magnitude with the $\Delta(k)$ values of (VII,11) as expected.

---

* For instance these neglected phase space factors contain that the formation of a symmetric deformation is more probable than the formation of an asymmetric deformation.

If one makes a more realistic assumption about the nuclear part of the cluster-interaction potential, for instance a rounded off potential well, then $\Delta E$ in (VII,18) will decrease because $R_2 - R_1$ increases, but the order of magnitude of $\Delta E$ will not change.

From the just given consideration one might naively argue that if the excitation energy for induced fission at threshold is increased by about 6 MeV, then the fission valley should approximately vanish because symmetric fission should now take place without tunneling. Experimentally this is certainly not the case. Even for much higher excitation energies a partly filled up fission valley remains. The reason for this relatively slow increase of the peak to valley ratio is that with increasing excitation energy the intermediate nucleus can fission *(asymmetrically* and *symmetrically)* in more and more configurations where the internal excitation energy of the clusters is not completely transformed into deformation energy. If one takes this effect into account by a statistical thermodynamical consideration then one finds that the peak to valley ratio, in agreement with experiment, decreases much slower with increasing excitation energy than would follow from the more naive argument above*. This consideration also reduces the $\Delta E$ value given in (VII,18), but the order of magnitude for this value again remains.

In conclusion we want to stress that it is certainly very difficult to make refined quantitative calculations for the description of the fission process. For instance it is at the moment impossible to make more than order of magnitude estimations about the absolute value of fission cross-sections. But with the help of qualitative energetical arguments we can gain a refined overall understanding of the fission mechanism by including in the liquid drop model considerations of cluster structure effects. For a more detailed discussion of fission from the cluster viewpoint considered here, see [127].

### C. Coulomb energy effects

If we neglect the influence of Coulomb forces completely, the level spectra of mirror nuclei should be identical due to the charge independence of nuclear forces. If we now assume the Coulomb energy to be the same in all states (ground and excited states) of the same nucleus, the total spectrum of each nucleus is shifted up a constant amount $E_c$ equal to the Coulomb energy of that nucleus. Hence under this assumption mirror levels of two mirror nuclei will all differ by the same energy difference, $\Delta E_c = E_c$ *(nucleus 1)* $- E_c$ *(nucleus 2)*, and the level spacing of mirror nuclei should be exactly the same. Due to the long-range character of the Coulomb forces this assumption is rather well fulfilled.

---

* Certainly the absolute value of the fission probability increases very fast with increasing excitation energy, as for instance is seen from the ratio of fission events to evaporated neutron emissions as a function of bombarding energy, see for this for instance J. HUIZENGA et al. [166].

The small remaining differences in the spacing of mirror nuclei therefore come from the different charge distributions in mirror levels. As the charge distribution of a nucleus is directly related to the wave function which describes the state in question, one would expect that the cluster structure will affect the Coulomb energy to a considerable extent. By studying the Coulomb energy behaviour from the energy spectra of mirror nuclei one can hope to gain some information about the cluster structure of the levels, K. WILDERMUTH and Y. TANG [147]. We emphasize this possibility arises because the influence of the purely nuclear forces is the same in corresponding mirror levels.

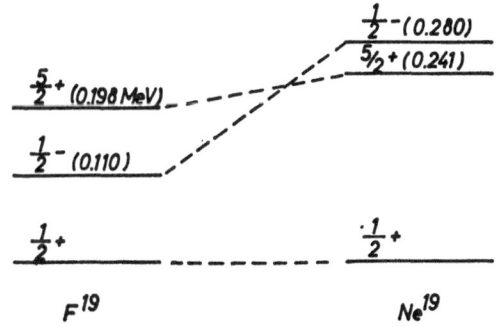

Fig. 29. Level diagram of F¹⁹ and Ne¹⁹

To see this in more detail, we consider as a special example the three lowest energy levels of the mirror pair $F^{19}$ and $Ne^{19}$ [148] (see Fig. 29). From detailed studies on the energy spectra it was concluded [149] that the cluster structure of the $\frac{1}{2}^+$ and $\frac{5}{2}^+$ levels of $F^{19}$ ($Ne^{19}$) is described by an unexcited $O^{16}$-cluster and a *triton* ($He^3$)-cluster in relative motion of orbital angular momentum $l = 0$ and $l = 2$. These two levels are therefore members of a rotational band, and we expect that the Coulomb energies should be approximately the same in both of these levels. Experimentally we indeed find that the excitation energies of the $\frac{5}{2}^+$ levels are nearly the same in both $F^{19}$ and $Ne^{19}$. For the $\frac{1}{2}^-$ levels, however, we have a different situation. These levels are best described by an unexcited $N^{15}$ ($O^{15}$)-cluster and an $\alpha$-cluster in relative oscillation of angular momentum $l = 0$. To obtain the configuration of these states from that of the ground states one moves a proton from the $O^{16}$ core in $F^{19}$ and a neutron from the $O^{16}$ core in $Ne^{19}$. Together with the outside *triton*- or $He^3$-cluster, an $\alpha$-cluster is then formed. This means that there is a decrease of Coulomb energy for $F^{19}$, but not for $Ne^{19}$, in first approximation. Therefore it is expected that the $\frac{1}{2}^-$ level of $F^{19}$ has a lesser excitation energy than the corresponding mirror level in $Ne^{19}$. Experimentally this is indeed the case (see Fig. 29).

10*

We can obtain a crude estimate of the difference in excitation energy $\Delta E$ of these $\frac{1}{2}^-$ mirror levels in $F^{19}$ and $Ne^{19}$ as follows. Since $\Delta E$ arises only from Coulomb energy differences we may write it formally exactly as

$$\Delta E \equiv E_{\text{exc}}\left(Ne^{19}, \frac{1}{2}^-\right) - E_{\text{exc}}\left(F^{19}, \frac{1}{2}^-\right)$$
$$= \left\{E_c\left(Ne^{19}, \frac{1}{2}^-\right) - E_c\left(Ne^{19}, \frac{1}{2}^+\right)\right\} \qquad \text{(VII,19)}$$
$$- \left\{E_c\left(F^{19}, \frac{1}{2}^-\right) - E_c\left(F^{19}, \frac{1}{2}^+\right)\right\}$$

where the Coulomb energies $E_c$ of the various levels are equal to the sum of the Coulomb self-energies of the belonging clusters plus their Coulomb interaction energy. Thus for the Coulomb energy of the $\frac{1}{2}^-$ level of $Ne^{19}$ we have, for example:

$$E_c\left(Ne^{19}, \frac{1}{2}^-\right) = E_c\left(O^{15}\text{-cluster}\right)$$
$$+ E_c\left(\alpha\text{-cluster}\right) + E_c\left(\text{interaction } O^{15}, \alpha\right).$$

To evaluate these Coulomb energies we now crudely assume the Coulomb self-energies of the clusters equal to the Coulomb energy of the corresponding free nuclei*. For the Coulomb interaction energies between the clusters in the various levels we equally crudely assume the charges of the $O^{16}$-, $N^{15}$- and $O^{15}$-clusters in the different levels to be distributed uniformly over a sphere of common radius $R_0$, and the charge of the $He^3$- and $t$-clusters, in their respective motions about the heavier clusters, to be smeared out uniformly over a larger sphere of radius $R_F$. The Coulomb interaction energy between the two clusters is taken to be the interaction energy between these two concentric uniformly charged spheres. With these assumptions we obtain from (VII,19)

$$\Delta E \approx \Delta E_c(O^{15}, N^{15}) - \Delta E_c(He^3, t)$$
$$- \frac{9\,e^2}{R_F}\left(1 - \frac{1}{5}\frac{R_0^2}{R_F^2}\right). \qquad \text{(VII,19a)}$$

The Coulomb energy difference of $O^{15}$ and $N^{15}$, $\Delta E_c(O^{15}, N^{15})$ and of $He^3$ and the triton, $\Delta E_c(He^3, t)$, are taken from the corresponding experimental values. $R_F$ and $R_0$ are taken as the radius of $F^{19}$ and $O^{16}$ respectively**. $R_F$ for instance may be evaluted from the Coulomb energy difference between the ground states of $Ne^{19}$ and $F^{19}$. With these values $\Delta E$ turns out to be 0.21 MeV; experimentally it is 0.17 MeV.

For all mirror levels whose cluster structures are known, one can calculate in a similar way the above mentioned excitation energy differences $\Delta E$. In Table 1 we list a number of such examples. It is

---

* This means we neglect all antisymmetrization effects in the calculation of the different Coulomb energies. Due to the long range character of the Coulomb force this is approximately justified.

** $R_0$ is taken from [92].

Table 1. *Excitation Energy Differences of Mirror Levels*

| Nuclear levels | Excitation energy (MeV) | Cluster structure (a) | Cluster structure of ground state | $\Delta E_{exp}$ (MeV) | $\Delta E_{theor}$ (MeV) | Remarks |
|---|---|---|---|---|---|---|
| $Li^5(He^5)$, $\frac{3}{2}^+$ | 16.81 (16.69) | $He^3(H^3)$—d, $2s$ | $\alpha$—p(n), $1p$ | + 0.12 | + 0.12 | b) |
| $Be^7(Li^7)$, $\frac{1}{2}^-$ | 0.431 (0.478) | $He^3(H^3)$—$\alpha$, $2p$ | $He^3(H^3)$—$\alpha$, $2p$ | — 0.05 | $\approx 0$ | c) |
| $Be^7(Li^7)$, $\frac{7}{2}^-$ | 4.54 (4.63) | $He^3(H^3)$—$\alpha$, $1f$ | $He^3(H^3)$—$\alpha$, $2p$ | — 0.09 | — 0.06 | b),c) |
| $Be^7(Li^7)$, $\frac{5}{2}^-$ | 7.18 (7.47) | $Li^6$ g.s.—p(n), $1p$ | $He^3(H^3)$—$\alpha$, $2p$ | — 0.29 | — 0.36 | |
| $C^{11}(B^{11})$, $\frac{1}{2}^-$ | 1.99 (2.13) | $Be^8$ g.s.—$He^3(H^3)$, $2p$ | $Be^8$ g.s.—$He^3(H^3)$, $2p$ | —0.14 | $\approx 0$ | |
| $N^{13}(C^{13})$, $\frac{1}{2}^+$ | 2.37 (3.09) | $C^{12}$ g.s.—p(n), $2s$ | $C^{12}$ g.s.—p(n), $1p$ | — 0.72 | — 0.60 | d) |
| $Ne^{19}(F^{19})$, $\frac{1}{2}^-$ | 0.28 (0.11) | $O^{16}$ g.s.($N^{15}$ g.s.)—$\alpha$, $5s$ | $O^{16}$ g.s.—$He^3(H^3)$, $4s$ | + 0.17 | + 0.21 | |
| $Ne^{19}(F^{19})$, $\frac{5}{2}^+$ | 0.241 (0.197) | $O^{16}$ g.s.—$He^3(H^3)$, $3d$ | $O^{16}$ g.s.—$He^3(H^3)$, $4s$ | + 0.04 | $\approx 0$ | c) |

a) In this column we list the internal structures and the modes of relative motion. For instance the $\frac{3}{2}^+$ level of $He^5$ is described by a deuteron cluster and a triton cluster in relative $2s$ state motion.

b) For the calculation of the $\Delta E$'s of the levels, we have used the cluster model wave functions described in these lectures and references.

c) Rotational level.

d) The distance of the outside nucleon from the center of the nucleus in the $\frac{1}{2}^+$ state of $N^{13}(C^{13})$ has been assumed the same as for the outside nucleon in the corresponding $\frac{1}{2}^+$ level (0.871, 0.5 MeV) in $F^{17}$; the latter distance can be calculated from their Coulomb energy differences.

seen that the agreement between the calculated and experimental values is fairly good. In particular the sign of $\Delta E$ is always predicted correctly. For levels in the same rotational band as the ground state, the magnitude of $\Delta E$ is always smaller than 0.1 MeV. From our arguments presented above this should indeed be expected.

Recent experiments seem to indicate that the excitation energy of the $\frac{3}{2}^+$ level in Li$^5$ is 16.64 MeV, F. AJZENBERG and T. LAURITSEN [150], instead of 16.81 MeV previously listed, [148]. If this is indeed so, then $\Delta E$ would be negative in contradiction to our prediction from the assumed cluster structure of this level. As there is at present moment ample evidence favouring a deuteron plus a triton cluster configuration for this state, see for instance [37] and chapter VI section B and C, we expect that this apparent discrepancy comes from the fact that it is rather difficult to determine to a good accuracy the energies of the ground states of He$^5$ and Li$^5$ because of the extreme widths of these states.

Finally, we wish to mention that one can also go the other way, that is, from the $\Delta E$ value of mirror levels one can also learn about the cluster structure of nuclear levels. For example let us consider briefly the $\frac{1}{2}^-$ levels of the mirror pair F$^{17}$ and O$^{17}$ at excitation energies of 3.10 and 3.06 MeV [148] respectively (see Fig. 30).

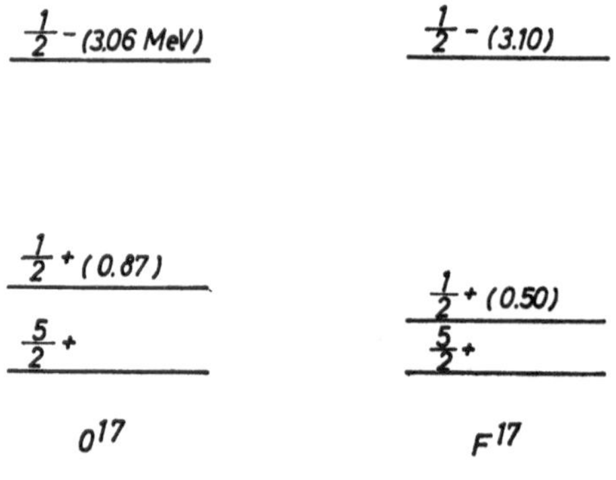

$\frac{1}{2}^-$ (3.06 MeV)                    $\frac{1}{2}^-$ (3.10)

$\frac{1}{2}^+$ (0.87)

$\frac{5}{2}^+$                                   $\frac{1}{2}^+$ (0.50)

                                              $\frac{5}{2}^+$

O$^{17}$                                        F$^{17}$

Fig. 30. Level diagram of O$^{17}$ and F$^{17}$

The change of parity in going from the ground state to the $\frac{1}{2}^-$ state requires in the shell model picture that one of the nucleons be raised to a higher shell. To obtain this state one could try exciting the odd nucleon outside the closed $1p$ shell from the $1d$ shell to the $2p$ shell. This would give the proper total spin (spin and orbital momentum antiparallel to each other) and parity of the level. In the case of O$^{17}$ it would be the odd neutron that would be moved further away from the nucleus, and there would be no Coulomb energy change. However in the case of F$^{17}$ it would be a proton which would be excited to the next higher shell, giving a relatively large decrease of Coulomb energy (about 0.5 MeV). By this reasoning we would expect the excitation energy of the $\frac{1}{2}^-$ level of F$^{17}$ to be about 0.5 MeV lower than the excitation energy of the O$^{17}$ mirror level. However experimentally one does not find this effect; indeed the excitation energy of the $\frac{1}{2}^-$ level of F$^{17}$ is even a little larger than the excitation energy of the O$^{17}$ mirror level*.

This shows that the structure described above is not the correct structure of these levels, and therefore one must look for another structure. One has to assume for the cluster structure of these mirror levels essentially an $\alpha$-cluster excitation of the O$^{16}$-cluster core. Although the lowest excitation of this kind in O$^{16}$ needs an energy of 6.06 MeV (see chapter III section E), one can account for the decrease of the excitation energy in F$^{17}$ and O$^{17}$ by the fact that the odd nucleon can now be in the $1p$ shell, which is left partly empty by the O$^{16}$ excitation. The cluster structure of O$^{17}$ is therefore now that of a C$^{13}$-cluster in the ground state plus an $\alpha$-cluster. Similar considerations give for the $\frac{1}{2}^+$ level in F$^{17}$ an N$^{13}$-cluster in the ground state plus an $\alpha$-cluster. There is a large spatial overlap of clusters in both cases. However in the case of F$^{17}$ we are moving a proton to an inner shell and increasing the Coulomb energy. The excitation energy of the $\frac{1}{2}^-$ level in F$^{17}$ should therefore be slightly larger than that of the $\frac{1}{2}^-$ level in O$^{17}$. This is also indicated by the experimental data. Furthermore it was found, T. GREEN and R. MIDDLETON [151], that this level in O$^{17}$ does not exhibit stripping in the (d, p) reaction, thus further supporting our assumption about the structure of this level which we inferred from the study of the Coulomb energy behaviour.

In nuclear isobars other than mirror nuclei, the corresponding mirror levels are those with the same isobaric spin quantum number $T$, etc. For example, in Li$^6$, He$^6$ and Be$^6$ the mirror levels are those with $T = 1$, [148]. Small differences in the level spacing can also give information about the cluster structure of these levels. There is no comparison for $T = 0$ levels since for these no mirror levels exist.

A further point we should mention is that rotational levels in a rotational band of a given cluster structure have all approximately the same charge distribution and hence the same Coulomb energy. This means that corresponding rotational bands, i.e. bands with mirror cluster structures, will have the same relative spacing of the rotational levels in both nuclei; however the distance of the rotational bands to the respective ground states of the mirror nuclei will not be the same if the ground states have a cluster structure different from that of the rotational bands, because different cluster structures in general have a different Coulomb energy. This is borne out by the experimental data, EVERLING [152].

---

* This tendency is even more clearly expressed if one considers the energy distances of the above mirror levels to the O$^{17}$ and F$^{17}$ $\frac{1}{2}^+$ mirror levels rather than the $\frac{5}{2}^+$ mirror levels.

## D. Reduced widths of nuclear levels

In the last section we discussed how the influence of the Coulomb interaction on the energy spectra of nuclear levels can be used to obtain valuable information about the structure of these levels. The disadvantage of the method is that it is restricted to mirror levels. Therefore we will discuss here two other methods which also can give information about the cluster structure of nuclear levels. The one is connected with the reduced widths of nuclear levels and the other with the $\gamma$-transition probabilities between nuclear levels, B. ROTH and K. WILDERMUTH [153].

We start with the discussion of reduced widths. An isolated nuclear resonance of a compound nucleus has an energy width $\Delta E = \Gamma$. To obtain a width which is more closely related to the nuclear structure one defines the reduced width by

$$\Delta E = \Gamma = 2k\,P\,\gamma^2 \qquad\qquad (\text{VII},20)$$

where $k$ is the relative wave number of the incoming particle in the center of mass system and $P$ the penetrability of the incoming particle into the target nucleus; $P$ is equal to unity for a high energy uncharged incoming particle of angular momentum zero and falls with increasing charge and angular momentum or with decreasing energy. The reduced width $\gamma$ can be expressed in dimensionless form as

$$\gamma^2 = \Theta^2\,\frac{3\,\hbar^2}{2\,\mu\,R} \qquad\qquad (\text{VII},21)$$

where $R$ is the radius, $\mu$ is the reduced mass for the relative motion, and $\Theta^2 \leq 1$. For $\Theta = 1$ we have the maximum reduced width $3\hbar^2/2\mu R$, called the Wigner limit, G. BREIT [1954], E. WIGNER and L. EISENBUD [155], see also E. VOGT [156] and [62].

We can obtain an estimate of this limit for the case of a high-energy incoming particle for which $P$ must approach unity, see [33] p. 104. Due to the uncertainty principle the width of the level $\Delta E$ is related to the time $\tau$ the cluster stays inside the target nucleus. Under the assumed conditions the particle will not be reflected back and forth within the nucleus, so we have $\tau = 2R/v$ where $v = \hbar\,k/\mu$ is the relative velocity.

We obtain therefore

$$\Delta E = \frac{\hbar}{\tau} = \frac{\hbar\,v}{2R} \qquad\qquad (\text{VII},22)$$

and from eq. (VII,20)*

$$\frac{\hbar}{2R} = 2k\,P\,\gamma^2 \approx \frac{2\,\mu\,v}{\hbar}\,\gamma^2\,. \qquad\qquad (\text{VII},23)$$

From this follows

$$\gamma^2 \approx \frac{\hbar^2}{4\,\mu\,R} \qquad\qquad (\text{VII},24)$$

which coincides, up to a factor of order of magnitude unity, with the Wigner limit.

---

* For the derivation of (VII,23) it is assumed that the velocity of the cluster is approximately the same inside and outside of the target nucleus. In the high energy limit case this is certainly correct within an order of magnitude.

To obtain the reduced width from the experimentally observed width we must always calculate the penetrability $P$. The usefulness of the reduced widths for the investigation of the cluster structures of nuclear levels comes about because they generally vary greatly enough with the change of the nuclear structure that moderate errors in the calculation of the penetrability will not obscure the qualitative information carried by the reduced widths.*

When the reduced width obtained from experimental data is of the same order of magnitude as the Wigner limit, we can assume that the square of the amplitude of the cluster structure corresponding to the reacting particles used in the experiment is close to unity. If other cluster structures have large amplitudes in the resonance state, the above reduced width will be much less than the Wigner limit and hence $\Theta^2$ will be much less than unity. This is so because the square root of the reduced width for a certain reaction channel $\lambda$ is given in general by, [154, 155],

$$\gamma_{\lambda o} \approx \int \psi_o \chi_\lambda \mathrm{d}s . \tag{VII,25}$$

Here $\psi_o$ is the true wave function of the resonance state of the compound nucleus at the nuclear surface; $\mathrm{d}s$ denotes the integration over the nuclear surface; and $\chi_\lambda$ is the wave function of the target nucleus and the bombarding particle at the nuclear surface, i.e. the wave function of the channel $\lambda$ being considered at the surface.

We shall illustrate the conclusions which may be drawn from the reduced widths by considering the case of $O^{16}$. The levels of $O^{16}$ are indicated in Fig. 31, [30, 150]. To the left of each level the cluster configuration believed to describe it is given in the oscillator model notation, and to the right the reduced width of the level for the scattering of $\alpha$-particles on unexcited $C^{12}$. The reduced widths are given as fractions of the Wigner limit. Because the $Q$-value for $\alpha + C^{12} \to O^{16}$ is 7.148 MeV, there are no reduced width data corresponding to the five lowest levels of $O^{16}$. The assignment of the cluster structure for many of these $O^{16}$ levels is discussed in chapter III section E.

The lowest of these levels which is energetically accessible directly by $\alpha - C^{12}$ scattering is the 8.87 MeV $2^-$ level. However, this level cannot be reached by the above scattering process because the $\alpha + C^{12}$ partial wave of angular momentum $l = 2$ has positive parity. Therefore its reduced width is zero.

The 9.58 MeV $1^-$ level should have a large reduced width, since it has the structure of an $\alpha$-cluster plus a $C^{12}$-cluster.

For the 9.84 MeV $2^+$ level the reduced width should be much smaller than the Wigner limit. This is due to the fact that this state is mainly described by the relative motion of an $\alpha$-cluster around an internally excited $C^{12}$-cluster.

---

* Because $P$ is rather poorly defined, the reduced width is a somewhat ambiguous notion theoretically. Complete antisymmetrization between target and incident particle leads to no essential difference from the usual approximate significance of reduced widths, however, as should be clear from our remarks on optical potentials, chap. V sect. D.

The 10.36 MeV $4^+$ level is a rotational state in which an $\alpha$-cluster "rotates" around an unexcited $C^{12}$-cluster, as discussed in chapter III section D. Therefore the reduced width should be of the order of the Wigner limit.

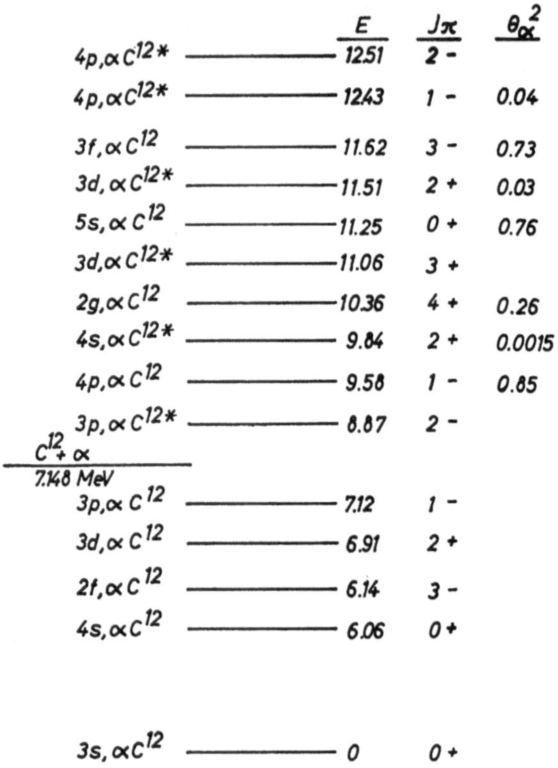

| | $E$ | $J\pi$ | $\theta_\alpha^2$ |
|---|---|---|---|
| $4p, \alpha C^{12}*$ | 12.51 | 2 - | |
| $4p, \alpha C^{12}*$ | 12.43 | 1 - | 0.04 |
| $3f, \alpha C^{12}$ | 11.62 | 3 - | 0.73 |
| $3d, \alpha C^{12}*$ | 11.51 | 2 + | 0.03 |
| $5s, \alpha C^{12}$ | 11.25 | 0 + | 0.76 |
| $3d, \alpha C^{12}*$ | 11.06 | 3 + | |
| $2g, \alpha C^{12}$ | 10.36 | 4 + | 0.26 |
| $4s, \alpha C^{12}*$ | 9.84 | 2 + | 0.0015 |
| $4p, \alpha C^{12}$ | 9.58 | 1 - | 0.85 |
| $3p, \alpha C^{12}*$ | 8.87 | 2 - | |
| $C^{12} + \alpha$   7.148 MeV | | | |
| $3p, \alpha C^{12}$ | 7.12 | 1 - | |
| $3d, \alpha C^{12}$ | 6.91 | 2 + | |
| $2f, \alpha C^{12}$ | 6.14 | 3 - | |
| $4s, \alpha C^{12}$ | 6.06 | 0 + | |
| $3s, \alpha C^{12}$ | 0 | 0 + | |

Fig. 31. Cluster structure and $\alpha$-particle reduced width for low lying $O^{16}$ levels

The 11.06 MeV $3^+$ level cannot be reached by $C^{12}(\alpha, \alpha) C^{12}$ scattering since the angular momentum $l = 3$ partial wave has negative parity. Therefore its reduced width is zero.

In a similar way the 11.25 MeV $0^+$ and the 11.623 MeV $3^-$ states should have large reduced widths, while the 11.51 MeV $2^+$ and the 12.43 MeV $1^-$ states should have small reduced widths because of their excited $C^{12}$-clusters.

It is seen that these predictions are confirmed in the measured values of the reduced widths. The 9.84 MeV $2^+$ level has a small reduced width for the scattering of $\alpha$-particles on unexcited $C^{12}$ nuclei, but it should have a large reduced width for the scattering of $\alpha$-particles on excited $C^{12}$ nuclei ($2^+$ state). In the latter case, in the calculation of

$$\gamma_{\lambda_0} \sim \int \psi_0 \chi_\lambda ds$$

there would be a large overlap of the scattering wave function at the nuclear surface with the compound nucleus state.

It is interesting to note that the proton reduced widths for the 12.43 and 12.51 MeV states are only 4 % of the Wigner limit, [30, 150]. These relatively small values are to be expected from the cluster structure of these states. The 12.51 state has in fact been found (KRAUS et al. [157], W. HORNYAK and R. SHERR [158]) to decay into an $\alpha$ and a $C^{12*}$ as well as decaying by $\gamma$ and proton emission. This again is expected from the cluster structure of the states.

These arguments are successful only because the changes in reduced widths are so marked that rather crude calculations of the penetrabilities can be tolerated. Due to approximations in the theory and experimental errors the reduced values quoted above are not much better than order of magnitude estimates, but they still enable us to reach important qualitative conclusions.

## E. Gamma transition probabilities

We come now to the second method, namely the information about the cluster structure of nuclear levels which can be obtained from their $\gamma$ transition probabilities [153]. There exists a relation between the branching ratios for $\gamma$ emission and the cluster structure of the nuclear

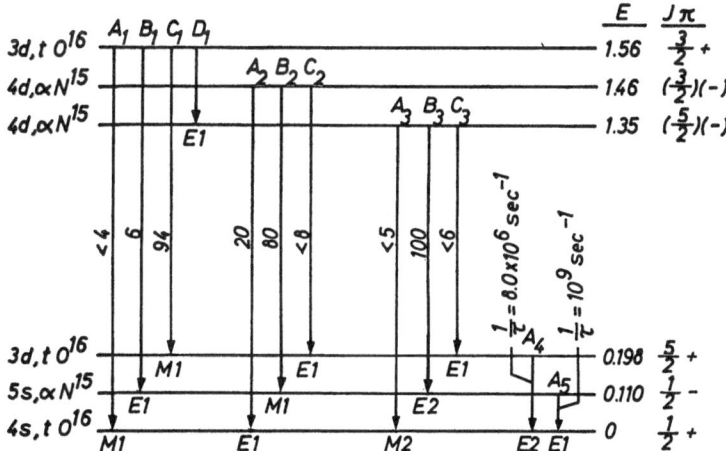

Fig. 32. Cluster structures and $\gamma$-transition probabilities for low-lying $F^{19}$ levels

states which also can be discussed qualitatively. This will be illustrated for the case of the relatively low lying $F^{19}$ states. The states of $F^{19}$, their cluster structures in the oscillator model notation and also the $\gamma$ transitions which we consider here are indicated in Fig. 32. The cluster structure of the ground state of $F^{19}$ is that of a *triton* cluster of spin $\frac{1}{2}$ in a *4s* oscillator state relative to the $O^{16}$-cluster. The 0.110 MeV $\frac{1}{2}^-$ level is that of an $\alpha$-cluster in a $5s$ oscillation about a $N^{15}$-cluster. The 0.198 MeV $\frac{5}{2}^+$ state has again a *triton* plus $O^{16}$-cluster structure and is the second rotational state (orbital angular momentum $l = 2$) in a rotational band whose lowest level is the ground state, see for this also [159 and 24]. The 1.35 MeV $\frac{5}{2}^-$ and the 1.46 MeV $\frac{3}{2}^-$ states have identical space configurations, that of an $\alpha$-cluster in $4d$ oscillation about a $N^{15}$-cluster, but in the $\frac{5}{2}^-$ state the spin direction of the $N^{15}$-cluster is parallel to the total $F^{19}$ orbital angular momentum direction and in the $\frac{3}{2}^-$ state these two directions are antiparallel to each other. The energy splitting of these two levels comes therefore from the spin-orbit coupling. The 1.56 MeV $\frac{3}{2}^+$ and 0.198 MeV $\frac{5}{2}^+$ states have also the same space configurations and opposite spin directions, this time of the *triton*-cluster spin relative to the total $F^{19}$ orbital angular momentum direction. But the spin-orbit energy splitting is much larger than in the case of the $\frac{3}{2}^-$ and $\frac{5}{2}^-$ levels, where the $N^{15}$-cluster has the internal spin $\frac{1}{2}$ and the relative orbital angular momentum of the $N^{15}$-cluster is therefore shared among fifteen particles, see for this [159].

Table 2 gives the transition probabilities for $\gamma$ transitions among these states, [150] and [30]. In the transitions from any particular level only the branching ratios of the various transitions are normally measured. Therefore in the transitions from the 1.56 MeV $\frac{3}{2}^+$ state we call the largest transition probability $p$ and express the other probabilities as fractions of $p$. The transition probabilities from the 1.46 MeV $\frac{3}{2}^-$ level are similarly defined in terms of $q$, and those from the 1.35 MeV $\frac{5}{2}^-$ level in terms of $r$. For

Table 2. *$\gamma$-transition probabilities in* $F^{19}$

| Transition | $w_{\exp}$ (sec.$^{-1}$) | $w_{\text{s.p.}}$ (sec.$^{-1}$) | $w_{\exp}/w_{\text{s.p.}}$ |
|---|---|---|---|
| $C_1$ | $p$ | $7.1 \times 10^{13}$ | $1.4 \times 10^{-14}p = |M(C_1)|^2$ |
| $D_1$ | $< 0.053p$ | $9.9 \times 10^{12}$ | $< 0.38 \cdot |M(C_1)|^2$ |
| $B_1$ | $0.064p$ | $3.3 \times 10^{15}$ | $1.4 \times 10^{-2} |M(C_1)|^2$ |
| $A_1$ | $< 0.043p$ | $1.1 \times 10^{14}$ | $< 2.8 \times 10^{-3} |M(C_1)|^2$ |
| $B_2$ | $q$ | $6.9 \times 10^{13}$ | $1.4 \times 10^{-14}q = |M(B_2)|^2$ |
| $A_2$ | $0.25q$ | $3.3 \times 10^{15}$ | $5.4 \times 10^{-3} |M(B_2)|^2$ |
| $C_2$ | $< 0.10q$ | $2.1 \times 10^{15}$ | $< 3.4 \times 10^{-3} |(M(B_2)|^2$ |
| $B_3$ | $r$ | $2.4 \times 10^{10}$ | $4.2 \times 10^{-11}r = |M(B_3)|^2$ |
| $A_3$ | $< 0.05r$ | $3.8 \times 10^9$ | $< 0.31 \cdot |M(B_3)|^2$ |
| $C_3$ | $< 0.06r$ | $1.6 \times 10^{15}$ | $< 0.9 \times 10^{-6} |M(B_3)|^2$ |
| $A_4$ | $8.0 \times 10^6$ | $2.6 \times 10^6$ | $3.1$ |
| $A_5$ | $1.0 \times 10^9$ | $1.4 \times 10^{12}$ | $0.72 \times 10^{-3}$ |

$w_{\exp}$ and $w_{\text{s.p.}}$ are the experimental and single particle transition probabilities the lowest levels in the table the absolute values of the transition probabilities are shown which are known from experiment.

The $w_{\text{s.p.}}$ are the single-particle $\gamma$ transition probabilities and can be calculated when the single-particle shell model notations of the initial and final states are known, J. BLATT and V. WEISSKOPF [160], S. MOSZKOWSKI [161], B. STECH [162]. If one assumes complete overlap of the single-particle nuclear wave functions in the initial and final states, the nuclear part of the transition matrix elements is equal to unity and one obtains for the electric and magnetic multipole transition probabilities:

$$T_E(l) = \frac{4.4\,(l+1)}{l\,([2l+1]!!)^2}\left(\frac{3}{l+3}\right)^2\left(\frac{\hbar\,\omega}{197\text{ MeV}}\right)^{2l+1}R^{2l}\cdot 10^{21}\,\text{sec}^{-1}$$

$$T_M(l) = \frac{1.9\,(l+1)}{l\,([2l+1]!!)^2}\left(\frac{3}{l+3}\right)^2\left(\frac{\hbar\,\omega}{197\text{ MeV}}\right)^{2l+1}R^{2l}\cdot 10^{21}\,\text{sec}^{-1}$$

$$(\text{VII,26})$$

where $R$ is the nuclear radius in units of $10^{-13}$ cm, and where $3!! = 3\cdot 1$, $4!! = 4\cdot 2$, $5!! = 5\cdot 3\cdot 1$, etc. These are the so-called Weisskopf single-particle transition probabilities. One sees that they are highly dependent on the energy difference in angular momentum between the initial and final states.

The above estimates can be refined if the nuclear wave functions are better known than in the single-particle shell model description, but for given transitions the above expressions are the definitions of the corresponding Weisskopf units.

We consider now the transitions $A_1$, $B_1$, $C_1$, and $D_1$. The $C_1$ transition is a transition from the $\frac{3}{2}^+$ to the $\frac{5}{2}^+$ level with $\Delta I = 1$ and no parity change. Therefore it is an $M_1$ (magnetic dipole) $\gamma$ transition and can only be a spin-flip transition. There is a large overlapping of the initial and final state due to their similar cluster structure, and the transition probability is therefore expected to be close to the corresponding Weisskopf unit, i.e. $\frac{W_{\text{exp}}}{W_{\text{s.p.}}} = |M(C_1)|^2 \approx 1$.

The $D_1$ transition is from a $\frac{3}{2}^+$ to a $\frac{5}{2}^-$ state and so must be an $E_1$ (electric dipole) transition. There is a difference in cluster structure between these two levels, the transition going from a *triton* plus $O^{16}$-clustering to an $\alpha$ plus $N^{15}$ clustering. There is therefore a relatively small overlapping in the structure of the initial and final state. In the $B_1$ transition one has the same difference in the cluster structure of the initial and final state as in the $D_1$ transition. Therefore the transition probabilities in both cases should be small compared with the corresponding Weisskopf units.

The $A_1$ transition is between two states of the same cluster structure. It is a $\Delta I = 1$ transition between a $\frac{3}{2}^+$ and $\frac{1}{2}^+$ state. However, it involves both an angular momentum change and a spin-flip. The magnetic dipole transition matrix element which enters into the first order calculation of the corresponding transition probability is therefore zero. This means one has to carry out a second order calculation to obtain this transition probability. This should give a very small magnetic dipole transition probability for the $\frac{3^+}{2} \to \frac{1^+}{2}$ transition compared with the corresponding Weisskopf value*.

In the second set of transitions the transition $B_2$ has the largest transition probability since it is a transition between states of the same cluster structure. The $B_2$ transition probability should be therefore close to the corresponding Weisskopf unit, i.e. $|M(B_2)|^2 \approx 1$. The transitions $A_2$ and $C_2$ involve changes in the cluster structure, and the corresponding transition probabilities expressed in Weisskopf units should therefore be small. Similar considerations can be made for the transitions $A_3$, $B_3$ and $C_3$. Here the electric quadrupole transition probability of the $B_3$ transition (no change in the cluster structure) should be close to its Weisskopf value $|M(B_3)|^2 \approx 1$. The other transition probabilities should be again much smaller than their Weisskopf units (change in the cluster structure).

In the $A_4$ transition one goes from the $\frac{5^+}{2}$ to the $\frac{1^+}{2}$ state with $\Delta I = 2$. There is no change in the cluster structure and we expect therefore again a transition probability close to its Weisskopf unit. One finds $W_{\text{exp}} = 8.0 \cdot 10^6 \text{ sec}^{-1}$ and $W_{\text{s.p.}} = 2.6 \cdot 10^6 \text{ sec}^{-1}$. Therefore $W_{\text{exp}}$ is equal to 3.1 Weisskopf units. The $A_5$ transition is between two states of different cluster descriptions and should therefore have a small value in Weisskopf units. Its value is found to be $0.71 \cdot 10^{-3}$ Weisskopf units.

The good order of magnitude agreement between experimental and predicted $\gamma$ transition ratios, as can be seen from Table 2, shows us the $\gamma$ transition probabilities between nuclear levels can also often be used to test cluster assignment of these levels. Again it is very important that these $\gamma$ transition ratios and probabilities change by orders of magnitude with changing cluster structures. This is the reason why our qualitative predictions work although we have not taken into consideration the indistinguishability of the nucleons.

It should be mentioned that similarly to the $\gamma$ transition probabilities the $\beta$-transition probabilities between nuclear states can be used sometimes to test the cluster assignment of these states, see for instance F. BARTIS [167].

## F. He³ and triton reactions in Fe⁵⁶

In the bombardment of $Fe^{56}$ with fast protons (energy of the protons between 100 MeV and 600 MeV) radiochemical methods reveal that besides many other reaction products, more than twice as many $He^3$ nuclei are produced as tritons, see for instance K. GOEBEL and J. ZAEHRINGER [163]. Because of the larger number of neutrons than protons, evaporation theory predicts 25 times more tritons than $He^3$ nuclei; this

---

* The electric quadrupole transition probability for the transition between these two states should be of the same order of magnitude or smaller.

is completely opposite to the experimental results. For the explanation of the experimental results there are two possibilities. First the production of He³ nuclei and tritons may be described as direct processes. This means in Fe⁵⁶ the incoming proton picks up first a proton and then a neutron, or first a neutron and then a proton, or first a neutron and then again a neutron. In this way one has two possibilites to produce a He³ nucleus and only one possibility to produce a triton. Therefore the direct interaction process would explain directly why more He³ nuclei than tritons are produced.

However, one can also explain the experimental results by the cluster structure of Fe⁵⁶, [24] pp. 116. The nucleus Fe⁵⁶ has 6 protons outside the closed neutron shell with the magic number 20 and 2 neutrons outside the closed neutron shell with the magic number 28. On the other hand one knows that Ca⁴⁸ is a stable isotope of Ca in spite of the large excess of neutrons, which means that the nuclear configuration $Z = 20$, $N = 28$ is very stable. Therefore we can assume that this configuration is not disturbed very much by the outside nucleons. We see now that with 6 protons and 2 neutrons as outside nucleons one can form either one *triton*-cluster with 5 remaining protons or two *He³*-clusters with two remaining protons. Without further explanation one sees directly that the second configuration is much more energetically favoured than the first. Therefore we expect that in the low energy states of F⁵⁶ the two *He³*-clusters plus 2 protons configuration is present with a much larger probability then the one *triton* plus 5 protons configuration. The excess of He³ nuclei over tritons in the reaction products is then explained either as a cluster evaporation effect or as a knock-out of the clusters by the incoming protons.

To see whether the direct effect or the cluster effect is more responsible for this He³ excess one can propose different test experiments. First Fe⁵⁶ is bombarded with high energy neutrons. If one then obtains a triton excess then the direct process is the more important process. But it is very difficult to do this experiment. Secondly one bombards Fe⁵⁶ with high energy deuterons. If one then obtains approximately the same number of He³ nuclei and tritons in the reaction products, then the direct process is more important. If essentially the old result remains, then the cluster effect is more important. Thirdly one bombards other nuclei having completely different cluster structures, as for instance Ca⁴⁰, Mg²⁴, etc., with high energy protons. If the relative He³ excess does not change essentially, then the direct process is the more important effect, otherwise the cluster effect. In practice this last experiment is the simplest of the three test experiments proposed.

Further, if the cluster effect is the more important, then one has to expect essentially low energy He³ nuclei and tritons ($E_{He³, t} \ll 50$ MeV) as reaction products; otherwise, high energy He³ nuclei and tritons ($E_{He³, t} > 50$ MeV) should be expected because in the pick-up process the outgoing He³ nucleus or triton receives nearly all the energy of the incoming proton. Therefore it would also be interesting to measure the energy distribution of the outgoing He³ nuclei and tritons.

With this we close our considerations in which we have qualitatively applied the physical concepts of cluster representations developed in the first six chapters to different examples. For other examples see [24] and [33].

For further discussion of cluster correlations, using group theoretical methods extensively, see V. NEUDACHIN and YU. SMIRNOV [171] and P. KRAMER and MOSHINSKY [172].

## VIII. Conclusions

In the preceding chapters we have developed the method of cluster representations of the atomic nuclei, which provides a unified approach to nuclear structure and reaction problems. The both qualitative and quantitative success of this approach stems from the fact that the nuclear forces are such as to favour certain correlations, to which have been applied the rather picturesque name clusters. It is often much more convenient both for purposes of physical insight and quantitative formulation to conceive of a nucleus in terms of a superposition of such cluster correlations, which can always be taken so as to form a complete if non-orthogonal set. By adjusting the amplitude of a cluster function in a given nuclear wave function we can more or less mix in as much of the corresponding cluster correlation as desired, and by varying various internal parameters we can regulate the strength of the correlation as desired.

Such a flexible representation requires an equally flexible quantitative approximation method for quantitative calculations, and we have seen that the Ritz variational procedure fulfills this need very well. With the Ritz principle we may include as many cluster terms as deemed necessary for the degree of approximation desired. In reaction problems this means we may independently open or close any channel we want to study or neglect, for in this method the only distinction between handling nuclear bound states and reactions is the application of proper boundary conditions to the appropriate cluster terms. We are thus enabled step by step to obtain increasingly better approximate solutions to the $A$-nucleon Schrödinger equation with just a few almost directly interpretable variational quantities. A corresponding variation in terms of the separate amplitudes of an arbitrary orthonormal set would usually require so many parameters as to be unmanageable. From the standpoint of such an arbitrary set, a set of cluster functions represents a select mixture of states in the proper ratio to describe a given cluster correlation, and we then handle this correlation as a whole rather than as a set of separate amplitudes.

We emphasize that any sort of physical effect can be included in the frame of the Ritz variational principle. For instance attention has been given to the importance of pairing correlations in nuclei, see for instance A. LANE [165]. One need only construct a trial nuclear wave function including terms incorporating such pairing correlations. The amplitudes of these terms may then be used as variational parameters, see for this also [165] p. 69. Similarly we may include Hartree-Fock terms in the trial function if there is reason to believe single-particle effects are particularly important for a given process. Such additional terms are usually not orthogonal to other terms in the trial wave function, but this is of no importance in a Ritz variational calculation.

In developing the cluster approach we have seen that the Pauli principle plays an extremely important role in determining the physical behaviour of nuclei. Perhaps the most important consequence is that under antisymmetrization, which is the quantitative expression of the Pauli principle, the difference between apparently different structures are greatly reduced. This is for instance the reason why apparently different nuclear models can be reconciled to one another without contradiction. In general if we start from a single-particle approach, antisymmetrization tends to reduce the single particle features and introduce some collective features, and if we start from a collective approach, antisymmetrization tends to reduce the collective features and introduce some single-particle features. Antisymmetrization also is the reason why clusters within a larger nucleus lose many of the properties of the corresponding free particles. It is the reason why in states where the clusters are highly overlapped, which is usually true in all low-lying nuclear levels, many cluster structures can exist at the same time with a high probability; for all these cluster structures differ only slightly in regard to the correlation properties of the nucleons. This last remark means that the method of cluster representations is in many respects a quantitative formulation of the compound nucleus considerations of N. BOHR [115], particularly if one regards the earlier resonating group formulation of the compound nucleus model by J. WHEELER [52]. At the same time it is also a generalization of the nuclear shell model of MAYER and JENSEN [48].

The small differences remaining between the various structures are nonetheless quite important. In nuclear reactions, for instance, there exist modes of excitation which affect only the relative motion of the clusters and leave the clusters themselves internally unexcited. With such excitations the cluster structure will not be changed, but the nucleon correlations within the clusters will become stronger as the separation distance between the clusters becomes larger until eventually the correlations go over into those of the corresponding free particles. However, because the differences in different cluster structures are usually small at low excitation energy, depending on the rearrangement of only a few nucleons, it is to be expected that very often the differences in excitation energy for these different structures will also be small, with the consequence that a slight change in the incident energy can lead into a completely different reaction channel. Using the same argument one can also explain why close lying bound states can possess entirely different cluster structures, as is experimentally confirmed for instance in the $F^{19}$ $\left(\frac{1}{2}^+\right)$ ground state, which has predominantly an $O^{16} - t$ cluster structure, and the $F^{19}$ $\left(\frac{1}{2}^{-}\right)$ state at 0.110 MeV, which has predominantly an $N^{15} - \alpha$ cluster structure, see chap. VII sect. C and E.

Finally we want to emphasize that the method of cluster representations is not a data-fitting model; it is a theoretical approach to solve approximately the $A$-nucleon Schrödinger equation. We employ no adjustable parameters once a realistic nuclear Hamiltonian has been chosen. The Hamiltonian for our $A$-nucleon system was taken to be a superposition of two-nucleon forces phenomenologically determined from the two-nucleon scattering data and deuteron bound state. This does not mean that three- or many-body forces are not in principle present (the answer to this question can only come from a more fundamental theory of the elementary particles), but rather that such forces are simply not important at the nucleon densities and energies found in nuclei. Nor are the quantitative results particularly sensitive to the form of the two-nucleon force itself so long as this interaction adequately describes the two-nucleon data*. This means that nuclear physics is not a particularly good tool for determining the exact form of the fundamental interaction between nucleons. More importantly it means that the method of cluster representations does not stand or fall on the exact form of the nuclear forces. For instance the inclusion of tensor forces, which are only poorly determined by the two-nucleon data and have been neglected in our discussions, will necessitate simply adding some more cluster terms in the nuclear wave function. Should in the future improved nuclear forces become available, the general approach to nuclear theory developed in the preceding chapters should not be altered beyond minor shifts in the numerical results.

### Appendix A

In this appendix we consider how the internal eigenstates of a cluster may be obtained for an internal Hamiltonian derived under the oscillator assumption. For definiteness we work with an $\alpha$-cluster, whose internal Hamiltonian is, see (III,8)

$$H_1 = \sum_{j=1}^{3} \sum_{k=1}^{3} \left\{ \left( \delta_{jk} - \frac{1}{4} \right) \frac{1}{2M} \, \overline{\boldsymbol{p}}_j \cdot \overline{\boldsymbol{p}}_k + (\delta_{jk} + 1) \frac{M \omega^2}{2} \, \overline{\boldsymbol{r}}_j \cdot \overline{\boldsymbol{r}}_k \right\} \quad (A,1)$$

where $\overline{\boldsymbol{p}}_j = \frac{\hbar}{i} \, \nabla_{\overline{\boldsymbol{r}}_j}$. The internal coordinates $\overline{\boldsymbol{r}}_j$ are defined in (III,6). To eliminate the cross terms in (A,1) we introduce as new cluster internal

---

    * For instance calculations of the $Be^8$ ground and first two excited states with a Yukawa potential do not show any remarkable differences from calculations with a Gaussian potential. See also [73] and [89].

coordinates the Jacobi coordinates:

$$z_1 \equiv r_1 - r_2 \qquad = \bar{r}_1 - \bar{r}_2$$

$$z_2 \equiv \frac{1}{2}(r_1 + r_2) - r_3 \qquad = \frac{1}{2}(\bar{r}_1 + \bar{r}_2) - \bar{r}_3$$

$$z_3 \equiv \frac{1}{3}(r_1 + r_2 + r_3) - r_4 = \frac{4}{3}(\bar{r}_1 + \bar{r}_2 + \bar{r}_3)$$

$$R_1 \equiv \frac{1}{4}(r_1 + r_2 + r_3 + r_4), \tag{A,2}$$

with inverse

$$r_1 = +\frac{1}{2}z_1 + \frac{1}{3}z_2 + \frac{1}{4}z_3 + R_1 = R_1 + \bar{r}_1$$

$$r_2 = -\frac{1}{2}z_1 + \frac{1}{3}z_2 + \frac{1}{4}z_3 + R_1 = R_1 + \bar{r}_2$$

$$r_3 = -\frac{2}{3}z_2 + \frac{1}{4}z_3 + R_1 \qquad = R_1 + \bar{r}_3$$

$$r_4 = -\frac{3}{4}z_3 + R_1. \tag{A,3}$$

It is often more convenient in practice to transform the single-particle oscillator Hamiltonian eq. (III,7) directly using eq. (A,2) and eq. (A,3), so for completeness we have included the transformation to the single-particle nucleon coordinates $r_i$, as well as to the old cluster internal coordinates, $\bar{r}_i$. If we let $q_j$ be the momentum canonical to $z_j$, then we have for the x-component of $\bar{p}_1$, say:

$$\bar{p}_{1x} \equiv \frac{\hbar}{i}\frac{\partial}{\partial \bar{r}_{1x}} = \sum_{j=1}^{3}\frac{\partial z_{jx}}{\partial \bar{r}_{1x}}\left(\frac{\hbar}{i}\frac{\partial}{\partial z_{jx}}\right) \equiv \sum_{j=1}^{3}\frac{\partial z_{jx}}{\partial \bar{r}_{1x}}q_{jx}$$

$$= q_{1x} + \frac{1}{2}q_{2x} + \frac{4}{3}q_{3x}. \tag{A,4}$$

Proceding similarly for all components we obtain the following relations between the canonical momenta of $\bar{r}_j$, $z_j$ and $r_j$:

$$\bar{p}_1 = +q_1 + \frac{1}{2}q_2 + \frac{4}{3}q_3 = p_1 - p_4$$

$$\bar{p}_2 = -q_1 + \frac{1}{2}q_2 + \frac{4}{3}q_3 = p_2 - p_4$$

$$\bar{p}_3 = -q_2 + \frac{4}{3}q_3 \qquad = p_3 - p_4$$

$$-q_3 + \frac{1}{4}P_1 \qquad = p_4 \tag{A,5}$$

where $P_1$ is the momentum canonical to the center of mass coordinate $R_1$, see eq. (III,5). Substituting (A,3) and (A,5) in (A,1), we obtain the transformed internal Hamiltonian

$$H_1 = \frac{1}{2M}\left(\frac{q_1^2}{1/2} + \frac{q_2^2}{2/3} + \frac{q_3^2}{3/4}\right)$$

$$+ \frac{M\omega^2}{2}\left(\frac{1}{2}z_1^2 + \frac{2}{3}z_2^2 + \frac{3}{4}z_3^2\right) \tag{A,6}$$

which now separates into a sum of oscillators of identical frequency $\omega$ and varying mass $\left(\dfrac{j}{j+1}\right) M$. The solution is then a product of oscillator functions. For an unexcited $\alpha$-cluster this is

$$\phi^0{}_{\text{space}} = \exp\left\{-\frac{a}{2}\left(\frac{1}{2}\,z_1^2 + \frac{2}{3}\,z_2^2 + \frac{3}{4}\,z_3^2\right)\right\}$$

$$= \exp\left\{-\frac{a}{2}\sum_{j=1}^{4}\bar{r}_j^2\right\} \tag{A,7}$$

where $\bar{r}_4 \equiv -(\bar{r}_1 + \bar{r}_2 + \bar{r}_3)$ as in (III,6) and $a = \dfrac{M\,\omega}{\hbar}$ is the width parameter. (A,7) agrees with our choice for $\phi^0{}_{\text{space}}$ in (III,18).

As long as we remain with the oscillator assumption the internal Hamiltonian of a cluster with any number of nucleons can always be brought to separable oscillator form by Jacobi coordinates. This is seen easily by induction. Thus suppose the single-particle oscillator Hamiltonian (which has the form (III,1)) for a cluster of $N$ nucleons in a larger nucleus has been brought to separable form by a Jacobi coordinate transformation:

$$\begin{cases} z_j = \dfrac{1}{j}\sum_{k=1}^{j} r_k - r_{j+1}\,; \quad j = 1, 2, \ldots, N-1 \\[2mm] R_1 = \dfrac{1}{N}\sum_{k=1}^{N} r_k \end{cases} \tag{A,8}$$

Now consider the addition of another nucleon to this cluster. The Hamiltonian for the new cluster of $N+1$ nucleons is then

$$H = \{H_1 + H_{\text{CM}}(1)\} + \left(\frac{p_{N+1}^2}{2M} + \frac{M\,\omega^2}{2}\,r_{N+1}^2\right)$$

$$= \left\{\sum_{j=1}^{N-1}\left(\frac{q_j^2}{2\left(\dfrac{j}{j+1}\right)M} + \left(\frac{j}{j+1}\right)\frac{M\,\omega^2}{2}\,z_j^2\right) + \frac{P_1^2}{2NM} + \frac{NM\,\omega^2}{2}\,R_1^2\right\} +$$

$$+ \left(\frac{p_{N+1}^2}{2M} + \frac{M\,\omega^2}{2}\,r_{N+1}^2\right). \tag{A,9}$$

We introduce center of mass and internal coordinates for the new cluster, $z_j'$ and $R_1'$, defined in terms of the coordinates $z_j$ and $R_1$ of the old cluster by

$$\begin{cases} z_1' = z_1 \\ \cdots\cdots\cdots \\ z_{N-1}' = z_{N-1} \\ z_N' = R_1 - r_{N+1} \\ R_1' = \dfrac{1}{N+1}(NR_1 + r_{N+1}) \end{cases} \qquad \begin{cases} z_1 = z_1' \\ \cdots\cdots\cdots \\ z_{N-1} = z_{N-1}' \\ R_1 = R_1' + \dfrac{1}{N+1}\,z_N' \\ r_{N+1} = R_1' - \dfrac{N}{N+1}\,z_N'. \end{cases} \tag{A,10}$$

The corresponding canonical momenta can be computed as in (A,4). Clearly the transformation (A,10) affects only the last two terms in (A,9).

Substituting (A,10) in (A,9) we obtain

$$H = H'_1 + H'_{CM}(1)$$

$$= \sum_{j=1}^{N} \left( \frac{q'^2_j}{2\left(\frac{j}{j+1}\right)M} + \left(\frac{j}{j+1}\right)\frac{M\omega^2}{2}z^2_j \right) + \quad\quad (A,11)$$

$$+ \frac{P'^2_1}{2(N+1)M} + \frac{(N+1)M\omega^2}{2}R'^2_1.$$

Thus if the internal Hamiltonian of a cluster of $N$ nucleons can be made separable, so can that for $N+1$ nucleons. Since we have already demonstrated the validity of (A,9) for four nucleons (2 and 3 nucleons are trivial to show), by induction (A,11) is valid for a cluster of any number of nucleons.

## Appendix B

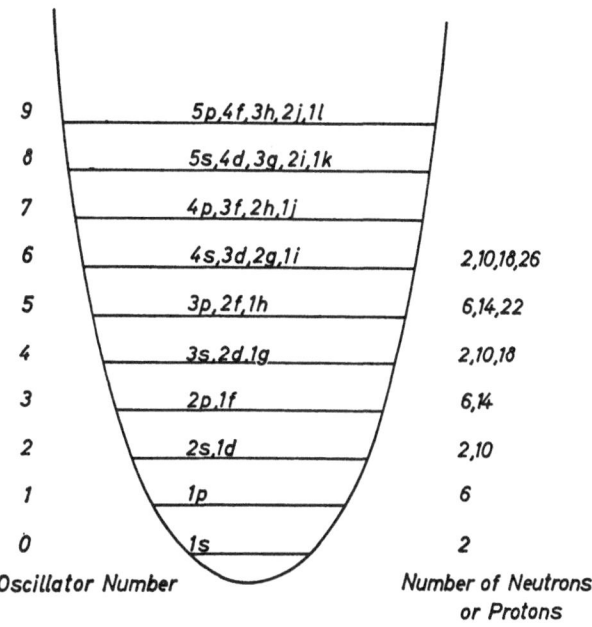

Notation in the oscillator model

## Appendix C

In this appendix we will show explicitly the mathematical equivalence in the oscillator picture of the wave functions for $Li^6$ expressed as a *triton-* plus a $He^3$-cluster, and as an $\alpha$- plus a *deuteron*-cluster, [33].

In the t − He³ picture, the wave function may be expressed as

$$\psi_{\text{Li}^\bullet} = \psi_{\text{I}} = \mathscr{A}\,\phi_0(t)\,\phi_0(He^3)\,\chi(\mathbf{R}_{\text{He}^\bullet} - \mathbf{R}_t) \tag{C,1}$$

where $\mathscr{A}$ is the antisymmetrizer, $\phi_0$ denotes the ground state internal wave function of a given cluster, and $\chi$ is the function describing the oscillatory relative motion of the two clusters. Similarly in the $\alpha$-deuteron picture the wave function may be expressed as

$$\psi_{\text{Li}^\bullet} = \psi_{\text{II}} = \mathscr{A}\,\phi_0(\alpha)\,\phi_0(d)\,\chi(\mathbf{R}_\alpha - \mathbf{R}_d)\;. \tag{C,2}$$

In these and the following equations the spin and isobaric spin part of the wave functions will only be written explicitly if needed for our considerations.

The functions $\chi(\mathbf{R}_{\text{He}^\bullet} - \mathbf{R}_t)$ and $\chi(\mathbf{R}_\alpha - \mathbf{R}_d)$ are taken to be in the lowest oscillator states not vanishing under antisymmetrization, namely in oscillator states* with oscillator quantum number 2. For this oscillator quantum number exist the 4 states $I = 1^+, 3^+, 2^+, 1^+$, as we have seen in chapter III section E. The orbital angular momentum of the relative motion of the first state is $l = 0$, for the other states $l = 2$. For simplicity we select for our further considerations the $I = 3^+$ state, where the orbital angular momentum $l = 2$ and the spin $S = 1$ of the deuteron cluster are parallel to each other**. The oscillator wave function of second order for $l = 2$, $l_z = 2$ without the Gaussian factor has the following well known form:

$$\chi_{2\hbar\omega} = (x + iy)^2\,, \quad l = 2, l_z = 2 \tag{C,3}$$

With (C,3) the two wave functions (C,1) and (C,2) become

$$\psi_{\text{I}} = \mathscr{A}\,\prod_{i=1}^{6} \exp\left\{-\frac{a}{2}\,r_i^2\right\}\,[(x_{\text{He}^\bullet} - x_t) + i(y_{\text{He}^\bullet} - y_t)]^2\,\xi_{\text{spin}} \tag{C,4}$$

$$\psi_{\text{II}} = \mathscr{A}\,\prod_{i=1}^{6} \exp\left\{-\frac{a}{2}\,r_i^2\right\}\,[(x_\alpha - x_d) + i(y_\alpha - y_d)]^2\,\xi_{\text{spin}}\;. \tag{C,5}$$

As discussed in chapter III section D the exponential parts have combined to give a symmetric common factor. The spin and isobaric spin factor is taken in both cases as

$$\xi_{\text{spin}} = \binom{1}{0}_{s_1} \binom{1}{0}_{t_1} \binom{0}{1}_{s_2} \binom{1}{0}_{t_2} \binom{1}{0}_{s_3} \binom{0}{1}_{t_3} \binom{0}{1}_{s_4} \binom{0}{1}_{t_4} \binom{1}{0}_{s_5} \binom{1}{0}_{t_5} \binom{1}{0}_{s_6} \binom{0}{1}_{t_6} \tag{C,6}$$

Upon antisymmetrization $\xi_{\text{spin}}$ becomes an $S = 1$, $T = 0$ eigenfunction.

---

* We are of course working in the simple oscillator picture, where the oscillator frequences of the cluster internal and relative functions are all identical.

** The proof of the equivalence of the above two cluster representations for the other Li⁶ states goes completely analogously to the $I = 3^+$ case considered here, only the calculations become a little more complicated.

The relative motion terms in (C,4) and (C,5) can be written as

$$[(x_{\text{He}^3} - x_t) + i(y_{\text{He}^3} - y_t)]^2$$

$$= \left[\frac{1}{3}\{(x_1 + i\,y_1) + (x_2 + i\,y_2) + (x_3 + i\,y_3)\} - \right. \tag{C,7}$$

$$\left. - \frac{1}{3}\{(x_4 + i\,y_4) + (x_5 + i\,y_5) + (x_6 + i\,y_6)\}\right]^2$$

$$[(x_\alpha - x_d) + i(y_\alpha - y_d)]^2$$

$$= \left[\frac{1}{4}\{(x_1 + i\,y_1) + (x_2 + i\,y_2) + (x_3 + i\,y_3) + (x_4 + i\,y_4)\} - \right. \tag{C,8}$$

$$\left. - \frac{1}{2}\{(x_5 + i\,y_5) + (x_6 + i\,y_6)\}\right]^2 .$$

As in the case of antisymmetrization of the $l = 4$, $l_z = 4$, $\text{Be}^8$ wave function in chapter III section D only such terms in the relative motion functions (C,7) and (C,8) will remain after antisymmetrization which describe that a proton and a neutron with parallel spins are in the $1p$ shell, i.e. terms of the form $(x_r + i\,y_r)(x_s + i\,y_s)$ with $r \neq s$. All these terms multiplied with the belonging spin function and the exponential part are proportional to each other after antisymmetrization*. With this we have proved that $\psi_{\text{I}}$ and $\psi_{\text{II}}$ are equal to each other up to constant factor.

# References

[1] HEISENBERG, W.: Z. Physik **77**, 1 (1932)
[2] IVANENKO, D.: Nature **129**, 798 (1932)
[3] BETHE, H. A.: Congrès International de Physique nucléaire, Paris 1964, p. p. 102—118, and following discussion
[4] BREIT, G., M. H. HULL JR., K. E. LASSILA, and K. D. PYATT JR.: Phys. Rev. **120**, 2227 (1960); **128**, 826 (1962)
    HULL, M. H. JR., K. E. LASSILA, H. M. RUPPEL, F. A. McDONALD, and G. BREIT: Phys. Rev. **128**, 830 (1962)
[5] WU, T., and T. OHMURA: Quantum Theory of Scattering. London: Prentice Hall 1962
[6] SCHMID, E.: Habilitationsschrift, T. H. München 1965
[7] MOSZKOWSKI, S. A.: Encyclopedia of Physics, Vol. 39, p. 464, Berlin-Göttingen-Heidelberg: Springer-Verlag 1957
[8] WILDERMUTH, K., and TH. KANELLOPOULOS: Nuclear Phys. **7**, 150 (1958)
[9] ELLIOT, J. P., and T. H. R. SKYRME: Proc. Roy. Soc. London A **232**, 561 (1955)
[10] BRINK, D. M.: Nuclear Phys. **4**, 215 (1957)
[11] LEDERER, K.: Diplomarbeit, München (1957)
[12] HEISENBERG-MACKE: Theorie des Atomkerns. Göttingen: Max-Planck-Inst., (1952) p. 72
[13] BRUECKNER, K.: Phys. Rev. **96**, 508 (1954), and later papers. See also: EDEN, R.: Progr. in Nuclear Phys. **6**, 26 (1957)

---

* For the more detailed proof see chapter III section D.

[14] GOMES, L. C., J. D. WALECKA, and V. F. WEISSKOPF: Ann. Phys. N. Y. **3**, 241 (1958)

[15] de SHALIT, A., and V. F. WEISSKOPF: Ann. Phys. N. Y. **5**, 282 (1958)

[16] BRENIG, W.: Nuclear Phys. **4**, 363 (1957)

[17] SCHLOEGL, F.: Z. Physik **136**, 445 (1953)
See also WIGNER, E. P.: Phys. Rev. **43**, 252 (1933); Z. Physik **83**, 253 (1933)

[18] WILD, W., and K. WILDERMUTH: Z. Naturforsch. **9**a, 799 (1954).

[19] CARLSON, B. C., and I. TALMI: Phys. Rev. **96**, 436 (1954)

[20] KURATH, D.: Phys. Rev. **101**, 216 (1956). See also FEENBERG, E., and E. WIGNER: Phys. Rev. **51**, 95 (1937), and WIGNER, E. P.: Phys. Rev. **51**, 106 and 947 (1937)

[21] WILDERMUTH, K., and TH. KANELLOPOULOS: Nuclear Phys. **9**, 449 (1958/59)

[22] FEENBERG, E., and M. PHILLIPS: Phys. Rev. **51**, 597 (1937)

[23] RACAH, G.: L. Farkas Memorial Volume (1952)

[24] WILDERMUTH, K., and TH. KANELLOPOULOS: CERN-Report 59—23 (1959)

[25] KANELLOPOULOS, TH., and K. WILDERMUTH: Nuclear Phys. **14**, 349 (1959)

[26] INGLIS, D. R.: Phys. Rev. **56**, 1175 (1939)

[27] PHILLIPS, G. C., and T. TOMBRELLO: Nuclear Phys. **19**, 555 (1960), and Phys. Rev. **122**, 224 (1961). SPENCER, R. G., G. C. PHILLIPS, and T. E. YOUNG: Nuclear Phys. **21**, 310 (1960)

[28] HOLMGREEN, H. D., and L. M. CAMERON: Rutherford Jubilee Conference, Manchester 531, 537 (1961). HOLMGREEN, H. D., and E. A. WOLICKI, Rutherford Jubilee Conf. 541

[29] TANG, Y. C., K. WILDERMUTH, and L. D. PEARLSTEIN: Phys. Rev. **123**, 548 (1961)

[30] AJZENBERG-SELOVE, F., and T. LAURITSEN: Nuclear Phys. **11**, 1 (1959)

[31] GALONSKY, A., and M. McELLISTREM: Phys. Rev. **98**, 590 (1955)

[32] JACKSON, D. F., and L. R. B. ELTON: Proc. Phys. Soc. London **85**, 659 (1965)

[33] WILDERMUTH, K.: Lectures on the cluster model of the nucleus: Technical Report No. 281, University of Maryland, p. 37 (1962); SHELINE, R. K., and K. WILDERMUTH: Nuclear Phys. **21**, 196 (1960)

[34] ELLIOT, J. P., and B. H. FLOWERS: Proc. Roy. Soc. London A **242**, 57 (1957)

[35] BROWN, G.: Unified theory of nuclear models. Amsterdam: North-Holland Publishing Company 1964

[36] PEARLSTEIN, L. D., Y. C. TANG, and K. WILDERMUTH: Phys. Rev. **120**, 224 (1960)

[37] KUNZ, W.: Phys. Rev. **97**, 456 (1955)

[38] TANG, Y. C., K. WILDERMUTH, and L. D. PEARLSTEIN: Nuclear Phys. **32**, 504 (1962)

[39] WILKINSON, D. H.: Rutherford Jubilee Conf. Manchester 339 (1961)

[40] WILDERMUTH, K.: Nuclear Phys. **31**, 478 (1962)

[41] —, and Y. C. TANG: Phys. Rev. Letters **6**, 17 (1961)

[42] PEARLSTEIN, L. D., Y. C. TANG, and K. WILDERMUTH: Nuclear Phys. **18**, 23 (1960)

[43] HACKENBROICH, H., G. WAHSWEILER, and K. WILDERMUTH: to be published

[44] See Ref. 12 p. 63

[45] MITTELSTAEDT, P.: Nuclear Phys. **17**, 499 (1960)

[46] See Ref. [14]

[47] See Ref. [15]

[48] GOEPPERT-MAYER, M., and J. H. D. JENSEN: Elementary theory of nuclear shell structure, p. 58. New York: JohnWiley and Sons 1955

[49] BOHR, A.: Kgl. Danske Videnskab. Selskab. Mat.-fys. Medd. **26**, No. 14 (1952)

[50] —, and B. R. MOTTELSON: Kgl. Danske Videnskab. Selskab, Mat.-fys. Med. **27**, No. 16 (1953)

[51] INGLIS, D. R.: Phys. Rev. **96**, 1059 (1954). But see also THOULESS D. J., and J. G. VALATIN: Phys. Rev. Letters **5**, 512 (1960), and Nuclear Phys. **31**, 211 (1962)

[52] WHEELER, J. A.: Phys. Rev. **52**, 1083 and 1107 (1937)

[53] PERRING, J. K., and T. H. R. SKYRME: Proc. Phys. Soc. London **69**, 600 (1956)
[54] PEIERLS, R. E., and J. YOCCOZ: Proc. Phys. Soc. London A **76**, 381 (1957)
[55] YOCCOZ, J.: Proc. Phys. Soc. London A **70**, 388 (1957)
[56] ALDER, K., A. BOHR, T. HUUS, B. MOTTELSON, and A. WINTHER: Rev. mod. Phys. **28**, 432 (1956), see in particular p. 454
[57] WIGNER, E. P.: Phys. Rev. **51**, 106 and 947 (1937)
[58] INGLIS, D. R.: Rev. mod. Phys. **25**, 390 (1953)
[59] ELLIOT, J. P., and B. H. FLOWERS: Proc. Roy. Soc. London A **229**, 536 (1955)
[60] ELLIOT, J. P.: Proc. Roy. Soc. London A **245**, 128 (1958)
     BAYMAN, B. F., and A. BOHR: Nuclear Phys. **9**, 596 (1958/59)
[61] VAN DER SPUY, E.: Nuclear Phys **11**, 615 (1959)
[62] LANE, A., and R. THOMAS: Rev. mod. Phys. **30**, 257 (1958)
     WATSON, K. M.: Phys. Rev. **88**, 1163 (1952)
[63] PHILLIPS, G. C., T. A. GRIFFY, and L. C. BIEDENHARN: Nuclear Phys. **21**, 327 (1960)
[64] BAZ, A.: J. Exptl. Theoret. Phys. U.S.S.R. **40**, 1511 (1961)
[65] PHILLIPS, G. C.: Rev. mod. Phys. **97**, 409 (1965)
     WAGGONER, M. A., J. E. ETTER, H. D. HOLMGREEN, and C. MOUZED: Rev. mod. Phys. **37**, 358 (1965)
     YOUNG, F. C., K. S. JAYRAMAN, J. E. ETTER, H. D. HOLMGREEN, and M. A. WAGGONER: Rev. mod. Phys. **37**, 362 (1965)
     DEHNHARD, D., D. KAMKE, and P. KRAMER: Phys. Letters **3**, 52 (1962)
[66] DAVIS, R. H.: Proc. of the Third Conf. on Reactions between Complex Nuclei held at Asilomar, Cal. (1963), remarks on page 49
     TAMURA, TARO: Rev. mod. Phys. **37**, 679 (1965)
[67] WHEELER, J. A.: Phys. Rev. **52**, 1083 and 1107 (1937)
[68] BUTCHER, A. C., and J. M. MCNAMEE: Proc. Phys. Soc. London A **74**, 329 (1959)
[69] SCHMID, E., and K. WILDERMUTH: Nuclear Phys. **26**, 463 (1961)
[70] OKAI, S., and S. C. PARK: to be published. See also SHIMODAYA, I., R. TAMAGAKI, and H. TANAKA: Progr. Theor. Phys. Kyoto **25**, 853 (1961), **27**, 793 (1962)
[71] HOFSTADTER, R.: Ann. Rev. Nuclear Sci. **7**, 231 (1957)
[72] HULTHÉN, L., and M. SUGAWARA: Encyclopedia of Physics. Vol. 39, 1. Berlin-Göttingen-Heidelberg: Springer-Verlag 1957
[73] TANG, Y. C., E. SCHMID, and K. WILDERMUTH: Phys. Rev. **131**, 2631 (1963)
[74] FOX, L., and E. GOODWIN: Proc. Cambridge Phil. Soc. **45**, 373 (1949)
[75] ROBERTSON, H. H.: Proc. Cambridge Phil. Soc. **52**, 538 (1956)
[76] SCHMID, E., Y. C. TANG, and K. WILDERMUTH: Proc. of the Conf. on Direct Interactions held at Padua 1962, pp. 62
[77] EDWARDS, S.: Proc. of the Conf. on Direct Interactions held at Padua 1962, pp. 469
[78] OKAI, S., S. C. PARK, and K. WILDERMUTH: Z. Physik **184**, 451 (1965)
[79] Proc. of the Conf. on Direct Interactions held at Padua 1962. See especially the article of R. SATCHLER pp. 80
[80] SCHMID, E., K. WILDERMUTH, and Y. C. TANG: Proc. of the Third Conference on Reactions between Complex Nuclei held at Asilomar Cal. (1963) pp. 59
[81] BENOEHR, H., and K. WILDERMUTH: to be published
[82] HOHLOCH, E., and K. WILDERMUTH: to be published
[83] WITTERN, H. W.: Nuclear Phys. **62**, 628 (1965)
     See also HARADA, M., R. TAMAGAKI, and H. TANAKA: Progr. Theor. Phys. Kyoto **29**, 933 (1963)
[84] BREIT, G., and E. P. WIGNER: Phys. Rev. **49**, 519 and 642 (1936)
[85] KAPUR, P., and R. PEIERLS: Proc. Roy. Soc. London A **166**, 277 (1938)
[86] FESHBACH, H.: Ann. Phys. N. Y. **5**, 357 (1958), and **19**, 287 (1962)
     See also SHAKIN, C.: Ann. Phys. N. Y. **22**, 54 (1963)

[87] HUMBLET, J., and L. ROSENFELD: Nuclear Phys. **26**, 529 (1961)
     ROSENFELD, C.: Nuclear Phys. **26**, 579 (1961)
[88] MACDONALD, W.: Nuclear Phys. **54**, 393 (1964); **56**, 636 and 647 (1964)
[89] SCHMID, E., Y. C. TANG, and R. C. HERNDON: Nuclear Phys. **42**, 95 (1963)
[90] WILDERMUTH, K.: Nuclear Phys. **31**, 478 (1962)
[91] JASTROW, R.: Phys. Rev. **98**, 1479 (1955)
[92] MEYER-BERKHOUT, U., K. W. FORD, and A. E. S. GREEN: Ann. Phys. N. Y. **8**, 119 (1959)
[93] BURLESON, G. R., and R. HOFSTADTER: Phys. Rev. **112**, 1282 (1958)
[94] JACKSON, D. F.: Proc. Phys. Soc. London **76**, 949 (1960)
[95] ELTON, L. R. B.: Nuclear Sizes. p. 21. Oxford: University Press 1961
[96] DABROWSKI, J., and J. SAVICKI: Acta Phys. Polon. **14**, 1323 (1955)
[97] VASHAKIDZE, I. SH., and G. A. CHITASHVILI: J. Exptl. Theoret. Phys. U.S.S.R. **29**, 157 (1955)
[98] KOPALEISHVILI, T. I., I. SH. VASHAKIDZE, V. I. MAMASHALHLISOV, and G. A. CHILASHVILI: Nuclear Phys. **23**, 430 (1961)
[99] SCHMID, E.: Nuclear Phys. **32**, 82 (1962)
[100] HANSTEEN, J. M., and H. W. WITTERN: Phys. Rev. **137**, B 524 (1965)
[101] OLLERHEAD, R. W., C. CHASMAN, and D. A. BROMLEY: Phys. Rev. **134**, B 74 (1964)
[102] ANDERSON, C. E., W. J. KNOX, and A. R. QUINTONY: Bull. Am. Phys. Soc. **5**, 292 (1960)
      — Reactions between Complex Nuclei, p. 67. New York: John Wiley and Sons, Inc. 1960
[103] HANSTEEN, J. M., and I. KANESTROEM: Nuclear Phys. **46**, 303 (1963)
[104] BOHR, N.: Kgl. Danske Videnskab. Selskab. Mat-fys. Medd. **18**, No 8 (1948)
[105] SCHMID, E., Y. C. TANG, and K. WILDERMUTH: Phys. Letters **7**, 263 (1963)
[106] FESHBACH, H., C. PORTER, and V. WEISSKOPF: Phys. Rev. **90**, 166 (1953); **96**, 448 (1954)
[107] NODVIK, J. S.: Nuclear Optical Model Conf. Tallahassee (1959)
[108] GOLDBERGER, M.: Phys. Rev. **74**, 1269 (1948)
[109] LANE, A. M., and C. F. WANDEL: Phys. Rev. **98**, 1554 (1955)
[110] MITTELSTAEDT, P.: Z. Naturforsch. **11**, 663 (1959)
[111] BRUECKNER, K.A., J. L. GAMEL, and H. WEITZNER: Phys.Rev.**110**,431 (1958), and earlier publications
[112] MELKANOFF, H. S.: Nuclear Optical Model Conf., Tallahassee (1959)
[113] WILDERMUTH, K., and R. L. CAROVILLANO: Nuclear Phys. **28**, 663 (1961)
[114] BLATT, J. M., and V. F. WEISSKOPF: Theoretical Nuclear Physics. New York: John Wiley and Sons 1952
[115] BOHR, N.: Nature **137**, 344 (1936)
[116] ALMQUIST, E., D. A. BROMLEY, and J. A. KUEHNER: Phys. Rev. Letters **4**, 365, 515 (1960)
[117] VOGT, E., and H. McMANUS: Phys. Rev. Letters **4**, 518 (1960)
[118] DAVIS, R. H.: Phys. Rev. Letters **4**, 521 (1960)
[119] WHITEHEAD and FOSTER: Can. J. Phys. **36**, 1276 (1958)
[120] SHERR, R., and collaborators: unpublished
[121] SCHMITT, H. W., and J. HALPERIN: Phys. Rev. **121**, 827 (1961)
[122] BOHR, N., and J. A. WHEELER: Phys. Rev. **56**, 426 (1939)
[123] FRENKEL, J. A.: J. Exptl. Theoret. Phys. U.S.S.R. **9**, 641; Phys. Rev. **55**, 987 (1939)
[124] FLÜGGE, S., and G. v. DROSTE: Z. physik. Chem. Leipzig (B) **42**, 274 (1939); Phys. Rev. **55**, 987 (1939)
      GOEPPERT-MAYER, M.: Phys. Rev. **74**, 235 (1948)
      WIEK, G. C.: Phys. Rev. **76**, 181 (1948)
      KOWARSKI, L.: Phys. Rev. **78**, 477 (1950)
[125] MEITNER, L.: Nature **165**, 561 (1950); Arkiv Fysik **4**, 983 (1952)
[126] FONG, P.: Phys. Rev. **102**, 434 (1956); Bull. Am. Phys. Soc. (II) **8**, 385 (1962); Phys. Rev. Letters **11**, 375 (1963)

[127] FAISSNER, H., and K. WILDERMUTH: Phys. Letters 2, 212 (1962); Nuclear Phys. 58, 177 (1964)
[128] HALPERN, I.: Ann. Rev. Nuclear Sci. 9, 245 (1959)
[129] HYDE, E. K.: OCRL 9036 (1960)
[130] FAIRHALL, A. W., R. C. JENSEN, and E. F. NEUZIL: Proc. 2nd Conf. Peaceful Uses Atomic Energy (United Nations, Geneva 1958) Vol. 15, p. 452
JENSEN, R. C., and A. W. FAIRHALL: Phys. Rev. 109, 942 (1958); Phys. Rev. 118, 771 (1960)
[131] BOHR, A.: Proc. 1st Int. Conf. Peaceful Uses Atomic Energy (United Nations, New York 1956) Vol. 2, p. 151
[132] HENKEL, R. L., and J. E. BROLLEY: Phys. Rev. 103, 1292 (1956)
[133] WHETSTONE, S. L.: Phys. Rev. 194, 581 (1959)
[134] VLADIMIRSKI, V. V.: J. Exptl. Theoret. Phys. U.S.S.R. 32, 822 (1957); J. Exptl. Theoret. Phys. 5, 673 (1957)
[135] HERRMANN, G.: Habilitationsschrift, Universität Mainz (1961), unpublished
[136] SCHLÖGL, F.: Z. Physik 136, 441 (1952)
[137] WILDERMUTH, K.: Fortschr. Physik 5, 469 (1957)
[138] VANDENBOSCH, R.: Nuclear Phys. 46, 129 (1963)
[139] SCHMITT, H. W., J. H. NEILER, F. J. WALTER, and A. CHETHAM-STRODE: Phys. Rev. Letters 9, 427 (1962)
[140] TEMMER, G.: private communication
[141] STEIN, W. E., and S. C. WHETSTONE: Phys. Rev. 110, 476 (1958)
[142] GIBSON, W. M., T. D. THOMAS, and G. L. MILLER: Phys. Rev. Letters 7, 65 (1961)
[143] MILTON, J. C. D., and J. S. FRASER: Phys. Rev. Letters 7, 67 (1961); Can. J. Phys. 40, 1626 (1963)
[144] JOHANNSEN, S. E. A.: Nuclear Phys. 12, 449 (1959); 22, 529 (1961)
[145] ARMBRUSTER, P., and H. MEISTER: Z. Physik 170, 274 (1962), and private communication
[146] GOEPPERT-MAYER, M., and J. H. D. JENSEN: Elementary Theory of Nuclear Shell Structure. p. 239 and 240. New York: John Wiley and Sons (1965)
[147] WILDERMUTH, K., and Y. C. TANG: Phys. Rev. Letters 6, 17 (1961)
[148] All the experimental data are taken from the compilation of F. AJZENBERG-SELOVE and T. LAURITSEN: Nuclear Phys. 11, 1 (1959)
[149] See for instance SHELINE, R. K., and K. WILDERMUTH: Nuclear Phys. 21, 196 (1960)
[150] See for instance T. LAURITSEN, and F. AJZENBERG-SELOVE: Energy Levels of Light Nuclei. Nuclear Data Sheets (May 1962)
[151] GREEN, T. S., and R. MIDDLETON: Proc. Phys. Soc. London A 69, 28 (1956)
[152] EVERLING, F.: private communication
[153] ROTH, B., and K. WILDERMUTH: Nuclear Phys. 20, 10 (1960)
[154] BREIT, G.: Theory of resonance reactions and allied topics. Vol. 41/1 Sect. 30. Encyclopedia of Physics, Berlin-Göttingen-Heidelberg: Springer-Verlag 1959
[155] WIGNER, E. P., and L. EISENBUD: Phys. Rev. 72, 29 (1947)
[156] VOGT, E.: in Nuclear Reactions, Vol. I, chapt. 5. Ed. ENDT and DEVNEUR. Amsterdam: North Holland Publishing Co., 1959
[157] KRAUS, FRENCH, FOWLER and LAURITSEN: Phys. Rev. 89, 299 (1953)
[158] HORNYAK, W., and R. SHERR: Phys. Rev. 100, 1409 (1955)
[159] SHELINE, R. K., and K. WILDERMUTH: Nuclear. Phys. 21, 196 (1960)
[160] BLATT, J., and V. F. WEISSKOPF: Theoretical Nucl. Phys. Chapt. 12. New York: John Wiley and Sons 1952
[161] MOSZKOWSKI, S. A.: in Beta and Gamma Ray Spectroscopy, chapt. 13. Ed. K. SIEGBAHN. Amsterdam: North-Holland Publishing Co., 1955
[162] STECH, B.: Z. Naturforsch. 7a, 401 (1952)
[163] GOEBEL, K., u. J. ZAEHRINGER: Z. Naturforsch. 16a, 231 (1961)
[164] ZELDÉS, N.: Nuclear Phys. 7, 27 (1958)
[165] LANE, A.: Nuclear Theory. New York, Amsterdam: W. A. Benjamin Inc. 1964
[166] HUIZENGA, J. R., R. CHAUDRY, and R. VANDENBOSCH: Phys. Rev. 126, 210 (1962)

[167] BARTIS, F. J.: Phys. Rev. **132**, 1763 (1963)
[168] TOMBRELLO, T. A., and P. D. PARKER: Phys. Rev. **130**, 1112 (1963)
[169] KHANNA, F. C., Y. C. TANG, and K. WILDERMUTH: Phys. Rev. **124**, 515 (1961)
[170] DAVIS, R.: Proc. of the 3rd Conf. on Reactions in Complex Nuclei. Berkeley: U. of Cal. Press (1963), p. 66
[171] NEUDACHIN, V. G., and YU. F. SMIRNOV: Atomic Energy Reviev, Vol. 3, No. 3, p. 151
[172] KRAMER, P., and M. MOSHINSKY: in Group Theory and its Applications, ed. by E. M. LOEBL. New York: Academic Press (to be published); see also Nuclear Phys. (to be published 1966)

Prof. KARL WILDERMUTH
Institut für Theoretische Physik der Universität
74 Tübingen, Gartenstraße 47
and
Dr. WALTER MCCLURE
Florida State University
Tallahassee, Fla.